I0823031

Diary of a
Honey Bee

Diary of a
Honey Bee

Dennis Wells

Foreword by Jürgen Tautz

Smithsonian Books
Washington, DC

Contents

To Yuma, Jesse and Lenny.
Stay curious.

Foreword

Our deep fascination with honey bees is no coincidence. Their products – honey and beeswax – have been treasured by humans for centuries. Since the groundbreaking work of Christian Konrad Sprengel in the eighteenth century, we have also understood the critical role bees play as pollinators of flowering plants. But admiration for bees stretches back even further. In ancient Egypt, their highly organized societies left such an impression that the honey bee became part of the royal hieroglyph as a prefix to the name of the pharaohs from the First Dynasty onwards. From the first microscopic preparations by Antoni van Leeuwenhoek in the seventeenth century to Stefan Diller's stunning high-tech images with the scanning electron microscope, bees have remained a source of fascination. Books about honey bees also look back on a long history.

Diary of a Honey Bee aims to tell a more personal story, revealing the lives of the extraordinary individuals in this infinitely fascinating world, and venturing into new territory in our understanding of honey bees as a species. One might think that understanding a bee colony as a whole means understanding the life of each individual worker bee, as they all seem the same. In fact, this is not the case. The misconception is understandable given that observing honey bees is a challenge, to say the least. As soon as they show up at the entrance of the hive, they take off and are gone. While observing the bustling activity of the bees inside a beehive, it

Bees have tiny hairs on their compound eyes. These hairs not only offer protection by preventing pollen getting stuck on the surface of the eye but also play a crucial role in navigation. Connected to a sensory cell, they enable bees to measure airflows while flying.

is impossible to distinguish individual bees from one another. The queen stands out because of her large body, and the drones, if present, due to their stocky build, but all worker bees look alike. However, simple colour markings can distinguish individual honey bees. Observations of them have revealed that, although each worker bee can perform every task in a beehive, the bees actually differ significantly, not only in terms of which tasks they perform but also in terms of how much work they do and even how well they do it.

We owe the first completely documented life history of a honey bee to the research of the bee scientist Martin Lindauer in the 1940s, who meticulously observed and documented the activities of individual bees throughout their entire lifespan. In the decades since Lindauer published his findings in the 1970s, bee research has become even more sophisticated. It has gradually become clear that each individual worker bee has its own personality, with characteristics and abilities that set it apart from the other bees in its colony. In principle, every worker bee is a jack-of-all-trades and can perform every duty in the colony. However, this does not mean that every bee actually does so. The closer you look, the more obvious the differences between the bees become. There are hard-working bees and lazy bees, early risers and late risers, sun lovers and a bad-weather task force. You can find good and bad 'dancers' (see waggle dance p. 61), curious bees and indifferent bees, very smart bees and rather clueless individuals. No matter what trait is tested, significant differences can be found. This begs the question of where the differences between bees come from and whether they make sense in the context of their coexistence.

Genetic and behavioural studies of individual bees and entire colonies offer us some insight. The characteristics of every organism are shaped by the interplay between their genetic makeup and environmental influences. In honey bee colonies, this interaction is particularly fascinating. Each virgin queen mates with multiple drones during her nuptial flight, which means a single colony has many fathers, each contributing different genetic material. Environmental factors – such as the food available and the temperature of the brood nest – also play a role, influencing the development and traits of the offspring. The most obvious of these is lifespan: summer bees live for only a few weeks, while winter bees can survive several months. The reason for the difference is clear. Winter bees, emerging in autumn, must make it through the cold winter months, while summer bees can be replaced continuously by new, short-lived workers. Beyond this question of lifespan, the diversity within a colony offers another, subtler advantage. It allows the colony to respond flexibly to its environment and the tasks at hand. A mix of abilities and traits among bees ensures that not all individuals react the same way to external stimuli, signals and cues. If every bee were to perform the same tasks at the same time, the colony – as a

All honey bees produce honey to survive, although their tastes may vary; some prefer it sweeter than others.

superorganism – could not adequately adapt to challenges. It would either overreact or underreact, rarely striking the right balance. Colonies made up of highly similar worker bees, if they existed, would face significant disadvantages in the evolutionary race. They would struggle to handle the varied and complex demands of daily life, leaving them unable to compete with colonies that benefit from the ecological advantage of diversity. Thus, because every bee in a hive has slightly different characteristics, preferences and personalities, present-day bee colonies have evolved to respond to their environment in remarkably nuanced ways. Not all bees respond to the same stimulus in the same way. Instead, every bee has its own response threshold – for example, with regard to temperature regulation. Adding to the variabilities in behavioural responses, the elaborate communication between the bees ensures the colony's collective response is both efficient and finely tuned.

Diary of a Honey Bee immerses the reader in the lives of the bees, enabling them to witness the life of a few individuals, representing the other tens of thousands of worker bees that collectively make an amazing society.

Prof. Dr Jürgen Tautz
University of Würzburg

In spring the number of bees in a healthy hive increases. Eventually the increased density of bees in the hive crosses a threshold, causing the bees to 'swarm', which means that half the bees from the hive look for a new home.

Bees can take several minutes to emerge from the cell after metamorphosis, but they always succeed.

OVERLEAF: Getting caught in the rain can be fatal for bees since they breathe through tiny openings in their exoskeleton.

The pristine 'Great Maple Floor' in the Karwendel mountain range, the oldest national park in Austria, is home to the maple tree where the bees in this book were followed for a full year.

Winter

In the winter

Honey bees are remarkably poorly adapted to winter. They are the only insects that die if their body temperature falls below 4°C (39.2°F). Many insect species produce a natural antifreeze that allows them to survive temperatures around or below freezing. Even as larvae, they can survive sub-zero temperatures with this trick. But honey bees cannot. They are extremely sensitive to low temperatures. After all, winter is a relatively new concept to them – at least in evolutionary terms. Honey bees originated in the tropical regions of Asia and Africa, first spreading to temperate latitudes a few million years ago. But they have developed a unique strategy to withstand the winters in temperate, continental climates that effectively keeps winter outside the hive.

The closer the better: the winter cluster

Honey bees are the only insects in temperate climates that remain active during winter. Most insects die at the end of autumn, and only their offspring survive the winter as eggs or larvae. With some insect species – such as wasps or bumble bees, which are closely related to honey bees – the colony dies and only the queen survives in a state of dormancy. Even ants lower their metabolism and activity levels, although they are relatively sheltered from the lowest temperatures in their underground nests. Some insects may come out of their hibernation if temperatures rise unexpectedly and be active for a short time, but as soon as temperatures drop, they go back to their inactive state. This phenomenon is known as a diapause.

But honey bees remain active throughout the winter by forming what is called a winter cluster. They cling together in a tight ball measuring about 30 centimetres (12 inches) in diameter. In a beehive, this ball is usually located directly below the lid, as that is the warmest spot (warm air rises). The surface temperature of the ball never drops below 10°C (50°F), while the centre of the ball is kept even warmer, at around 20°C (68°F). The bees take turns crawling to the centre to warm up briefly before returning to the colder edge of the winter cluster to perform their duty. The bees use their bodies, or their flight muscles to be exact, to generate the heat necessary for their winter cluster. Like all insects, bees are cold-blooded animals. This means their bodies are as warm or cold as the ambient temperature unless they use their muscles to generate heat. Honey bees are able to decouple their

A bee colony clustering together in a ball on a cold day in winter. In the centre the queen bee is protected from dangerously low temperatures.

wings from their flight muscles and use the muscles to vibrate, without moving their wings. They alternate contracting the muscles against one another so precisely that the movements cancel each other out and their vibrations are not perceptible to the human eye. Just like shivering humans, vibrating bees produce body heat and can raise their body temperature to over 43°C (109.4°F) and maintain that for about half an hour. Afterwards, they are completely exhausted and need to replenish their energy, handing over their heating responsibilities to other bees.

In a colony of 5,000 to 10,000 bees, there are enough individuals to take turns heating and maintaining the temperature in the cluster high enough for all the bees to survive in deep winter. This is critical for the most valuable member of the colony in particular – the queen bee – who stays at the centre of the winter cluster. Without her, there would be no colony; but without the colony, the low temperatures would be her death sentence. Together, however, they stand a good chance of surviving the winter.

Shaping their own world

The winter cluster is a prime example of how exceptional bees are. As bee expert Prof. Jürgen Tautz says, 'Bees shape their own world.' In winter, they generate the warmth they need to survive, while in summer, they produce the food their colony needs: honey. Their ability to build honeycombs is another great example as the bees produce the building material for the thousands of perfectly symmetrical brood cells themselves. In some ways, bees have managed to become independent of their environment.

Their collective achievements are incredible: tens of thousands of small individuals perfectly in sync. It is only because thousands of bees collect millions of droplets of nectar from millions of flowers that honey bees can make it through

The colder the environment, the closer the bees huddle together to stay warm. The queen is at the centre and therefore not visible.

winter. This is the reason why they make honey: to stockpile food for winter. In this regard – shaping their own living conditions and working together as a team – bees are not dissimilar to humans, albeit on a completely different scale. Each individual bee contributes significantly to the impressive feats of the collective.

In recent years, more and more scientists have shown that bees are not all identical, mindless members of a colony, performing the duties assigned to them by nature like robots. Instead, they are genuine individuals with different personalities, inclinations, and abilities based on their unique genetic makeup and learning experiences. These diverse personalities and experiences enable the bee colony to deal with any challenge it faces and adapt to changing environmental conditions. And the uniqueness and value of each individual bee's contribution to the performance of the collective makes their collaboration even more impressive.

Running hot or cold

A good example of how bees within a colony differ from one another, and how these differences benefit the entire colony, is how they heat the winter cluster. Bees vary in terms of how sensitive they are to heat or cold, with different 'favourite' temperature ranges. Some bees prefer cool temperatures, while others like it hot. Therefore, each bee starts and stops heating the winter cluster at a slightly different ambient temperature. Having a uniform upper or lower temperature limit would be much less efficient because it would mean that all the bees would start and stop heating simultaneously, only to start again shortly thereafter. As a result, they would consume more energy and have less time to dedicate to activities other than heating. If engineers were to build a heating system with thousands of small heaters, they would also ensure that they turned on and off at slightly different temperatures rather than all at the same time. Bees have been implementing this principle for millions of years.

The queen mother

The queen of a bee colony is surrounded by a court of about ten to twelve bees, which attend to her every need. The court always faces the queen, sometimes even forming a perfect circle around her. They feed her continuously and remove her waste. Additionally, they clean her regularly to ensure that no pathogens pose a threat to her, as that would mean the end of the colony. Although that is not entirely true – more on that later (see p. 100).

A queen is always surrounded by her 'court' of about twelve bees that keep her clean and fed. A queen can live for up to five years.

A winter bee drinking from a cell filled with honey collected by summer bees.

The court is in constant contact with the queen so they can communicate to the rest of the colony that she is well. She continuously emits a pheromone (the queen mandibular pheromone), which has a citrus-like scent that is too faint for humans to perceive but is clearly noticeable to bees. The court bees collect this scent with their tongues and spread it to all the bees in the colony through touch. Doing so reassures the tens of thousands of individuals in the colony that the queen is doing well.

How honey becomes liquid

The winter cluster formed by the bees extends across several combs. If one or more combs cut through the formation, this does not reduce the insulating effect. However, the bees also need to live off something during winter, and this is where honey comes into play. Honey bees produce honey (see p. 98) so they can live off it through the winter when there is no other food source available to them. But honey solidifies in winter after being stored in the combs over an extended period of time at temperatures below 25°C (77°F). In order to consume the crystallized honey, the bees first need to warm it up. Therefore, every three to four days, the bees indulge in a 'warm day', when they collectively heat up the winter cluster to a toasty 36°C (96.8°F) to return the honey to its liquid state. They then suck up the liquid honey with their proboscises. The proboscis is an anatomical wonder, tucked away when bees are in flight and only revealing itself to us when bees extend it to drink liquids. Unlike the proboscis of butterflies or mosquitoes, a bee's proboscis has a narrower, extendible tube inside it called the glossa. This functions like a tongue and is covered in over 10,000 hairs. This tongue moves up and down while feeding, enabling the bee to both lick and suck. Once the bees have used up all the honey reserves in one part of the comb, the winter cluster simply moves to another area that still contains honey.

Thus, the winter cluster moves across the comb, allowing all the bees to have a meal. This is critical as honey is the energy source the bees need to heat the cluster by vibrating their flight muscles. This finely tuned interplay of temperature, position of the winter cluster, and honey reserves can go awry. Even if there are still reserves of

The honey in the cells must sustain the winter bees for almost half a year. It also serves as 'fuel' to keep the winter bees warm.

Bee wings are an extremely lightweight yet stable structure. Small ridges ensure this stability and serve to stiffen the wings. All bees share the same pattern of ridges.

honey left in the hive, they might be out of reach for the bees in the winter cluster. A single bee cannot simply venture off into the hive to search for honey when temperatures are dangerously low. And even if she did dare to try, she would not be able to liquefy the honey on her own. So, even when huddled together in the winter cluster, a bee colony can starve to death if it does not move far enough across the comb to access all the honey reserves.

Eggs and larvae

As the days grow longer and temperatures begin to rise, the end of winter comes into sight. Deep inside the hive, the queen does not perceive the light of the lengthening days, but she can sense when the bees around her do not need to warm up as often because the temperatures outside are becoming milder and spring is approaching. At this point, she decides to lay the first egg and kick off a new bee year. The first egg is just one of tens of thousands that the queen will lay each year. During spring, she lays 2,000 eggs per day – an extraordinary feat considering the eggs she lays on each day together weigh as much as she does. Her body needs a lot of high-quality food to produce so many eggs. Each time she lays an egg, she also has to decide whether to fertilize it with sperm from her supply to make a female bee or else leave it unfertilized and create a male bee, a drone.

Since drones are not needed at this early stage of the year, she only lays fertilized eggs at first. From the moment an egg is deposited in a cell, the collective begins to care for it. Mould and bacteria are removed from the surface of the egg and the surrounding cell. Each egg is nurtured and cared for by the bees. After just three days, a larva only a few millimetres long hatches.

Feeding the larvae royal jelly

The tiny larva hatches into a veritable paradise. In its cell, it bathes in a temperature-controlled nutritious solution provided by the nurse bees. Glands in their heads produce this solution that they then secrete through other specialized glands next to their mouthparts. It is the counterpart to the 'mother's milk' of mammals, except

A queen laying eggs. The queen lays fertilized eggs in the smaller cells, which will develop into females, and unfertilized eggs in the larger cells, which will develop into males.

that it could be called 'sister's milk' because the queen, the actual mother, does not have time to care for each larva. Instead, the larva's older sisters perform this task. This 'sister's milk' has risen to fame under the name 'royal jelly'. It is undoubtedly an extremely nutritious liquid, rich in antibodies that boost the immune systems of the larvae. Whether it belongs in our cosmetics is a completely different question.

The milky solution is exclusively used to nourish the offspring. Bees need to consume considerable amounts of honey and pollen to produce it, which fortunately are readily available in the colony's hive (or nest). It is a widespread misconception that only queen bees are fed royal jelly. *All* bee larvae receive this 'sister's milk'. However, the 'sister's milk' for the queens is sweeter. More importantly, the queen larva is fed the milk throughout her time as a larva, whereas 'normal' bee larvae have to make do with ordinary 'bee bread' (a mixture of royal jelly, pollen, and honey) towards the end of the larval stage.

Nurse bees plunge deep into the cells to deposit royal jelly to feed the larvae. This 'sister's milk' is produced in special glands in the bees' heads.

A bee larva, a few days old. It feeds on the royal jelly surrounding it, which is produced by the nurse bees.

The uterus of the bees

Another surprising similarity between bees and mammals is the brood nest: the area in the centre of the comb where the queen lays her eggs. Prof. Jürgen Tautz calls it the 'uterus of the bees'. The queen first lays the eggs in the centre of the comb and then gradually moves outwards. Once the first egg has been laid, the bees do not allow the temperature of the brood nest to drop below 34°C (93.2°F), ideally keeping it at 36°C (96.8°F). Employing the same technique they use to drastically increase their body temperature in the winter cluster, they now heat the area of the comb with the larvae.

Heating and refuelling

Bees are one of the few insects that actively brood their offspring, and they do this at almost the same temperature range that birds use to incubate their eggs and that mammals have in their uterus: 36–37°C (96.8–98.6°F). To ensure the heat is distributed as evenly as possible among the larvae, the bees leave individual cells in the brood nest empty. They can crawl into these cells and heat them, with the resulting warmth spreading out better among the cells containing larvae than if the bees were to generate heat from on top of the cells.

Prof. Anton Stabentheiner from the University of Graz has shown that not all bees are equally good at heating. It seems some bees are particularly talented at this – and since some bees heat much more frequently and for longer periods than others, they can rightfully be called 'heater bees'. There are also 'refueller bees', which are primarily responsible for supplying the heater bees with honey so they do not have to leave their post. They fetch it from the honey cells and transport it in their crop – a kind of honey stomach – to the heater bees.

Bees regularly feed each other, sharing fresh nectar or honey among themselves. 'Refueller bees' are specialized in feeding other bees.

The reason bees put so much effort into regulating the temperature of the brood nest with such precision is linked to the next stage of development. After nine days, the bees stop feeding the larvae. By this point, they have grown so large that they almost completely fill their cells. The worker bees then seal the cells with a cap of beeswax, and the miracle of metamorphosis commences.

Metamorphosis: a full transformation

Now the organism undergoes a full transformation. In just 11 days, the body of the larva – a pupa with no limbs, eyes, or wings – morphs into an insect with six legs, five eyes (see p. 52), antennae, and, of course, wings. The internal organs of the bee and the chitinous exoskeleton also develop during this time. Metamorphosis may appear miraculous to us, but 80 per cent of all insects undergo it, which equates to well over half of all animal species on our planet. It is an ancient, very common, and extremely successful adaptation strategy. Contrary to earlier beliefs, the body of a larva does not completely dissolve into primordial soup before reforming. Rather, hormones control the radical restructuring of the body; all tissue types of the larva are fundamentally altered and repurposed. Many cells also die off and are replaced by stem cells, which can grow into entirely new cell types. The valuable contents of the dying cells are not wasted however; they are recycled to create new cells.

A bee 'capping' a cell with a larva inside. It seals the cell with wax, allowing the larva to metamorphose into a bee.

Delicate conditions

Temperature plays a key role in this complex process because optimal metamorphosis depends on a narrow range of a few degrees. Prof. Jürgen Tautz and his team have examined this temperature range more closely and found that bees transforming from a pupa (the stage after the larval stage) to a bee at only 34.5°C (94.1 °F) perform worse on intelligence tests than bees that undergo this process at the usual 36°C (96.8°F). At slightly lower temperatures, bees may not develop particularly well, but Prof. Tautz believes there could be more to it than that. Heating the brood nest requires energy, which is generated by using up the painstakingly produced honey (food for the heater bees). Bees use honey very sparingly because it is so laborious to produce, so bees might choose to generate less heat when they do not need particularly intelligent bees. For example, they may need skilled forager bees in the summer to fly from flower to flower and find their way back home, but they need them less when outside temperatures drop in the autumn and the colony just needs to be heated. So perhaps bees create the type of bees that are appropriate for that particular season. This could also affect not only their abilities in terms of being heater or forager bees but also their other inclinations and characteristics.

At long last: an adult bee emerges

Twenty-one days after the first egg was laid, and 12 days after the cell was capped, the bee that was a larva not long ago bites open the cell cap from the inside and begins to emerge. It is an exhausting process, and no-one helps, even though there are many other bees walking around nearby. She repeatedly tries to come out only to slip back inside and has to enlarge the opening using her mandibles. She uses her legs as we would use our elbows to free herself from the confinement of the cell and pushes herself out after a few minutes. And then she is there: the first bee of the year. Shortly after hatching, the other bees finally take notice and immediately inspect the new bee. One of the winter bees checks, for instance, if her wings are present. If her wings are deformed – for example, due to a viral infection – the freshly emerged bee is immediately thrown out of the nest to prevent her infecting other larvae. But if the new bee passes the inspection, she is accepted into the colony. Bees greet each other with their antennae – a highly sensitive organ. Each antenna possesses around 20,000 sensory cells. A bee can perceive temperature, humidity, smell, and taste with her antennae. A bee 'shakes hands' for the first time in her life when she touches antennae with another bee. As with humans, there are both right-handed and left-handed bees, or rather, bees that prefer to use their right or left antenna. Right-handed bees are significantly more common than left-handed bees.

Bees use their antennae to identify one another, mainly by smell. During the first few minutes after hatching, when the bee still has a silvery, moist coat, she is 'neutral', meaning she has not yet adopted the scent of the colony. If this young bee was placed in a different colony, she would be accepted there as well. But after a few minutes in her own colony, she has taken on its characteristic scent. This makes her an unmistakable member of that colony for life. She would not be accepted in another colony and would be forcibly expelled if she tried to enter it.

Not all honey bees are the same

Now the question arises: What kind of bee will she be? Will she be a diligent bee? Not all bees in the hive work equally hard. Will she be a particularly curious bee? Some bees fly farther than others. Will she be particularly skilled at building cells or take great care of the offspring? About a hundred years ago, the venerable naturalist and Nobel laureate Karl von Frisch noted that not all honey bees are alike. In 1934, he wrote about the bees' sense of taste, including a paragraph on 'individuality'. Von Frisch offered bees two different nutrient solutions a few metres away from the beehive. He mixed a few drops of a sour substance into one sugar solution and a few drops of a bitter substance into the other. Then, over a period of 24 days, he observed whether the bees would accept the bad-tasting sugar solutions at all. While some bees were not picky, other bees rejected solutions that they perceived as either too sour or

A bee emerging, a feat of strength that can take several minutes but does not require any help from other bees.

TOP LEFT: The bee gnaws at the wax cap from inside the cell.

MIDDLE LEFT: The emerging bee often slides back into the cell multiple times to enlarge the opening before fully emerging.

BOTTOM LEFT: A bee emerging, with remnants of the pupal skin still visible on her head.

TOP: The fur of a newly emerged bee is wet and silvery for a few minutes.

BOTTOM: After a short time, a new bee can hardly be distinguished from her peers.

too bitter. It was a matter of individual taste. One individual did not like any of the sugar solutions on offer.

More recently, Prof. Robert Page, an entomologist at Arizona State University, discovered that differences in sugar sensitivity are already evident in bees only a few hours old. It turns out that these preferences ultimately determine whether a bee will later become a pollen collector or a nectar collector. Not only that, but among the nectar collectors, some bees prefer certain flowers over others. Only in the case of an emergency – when they are urged to by other bees in the hive – will they also fly to flowers whose nectar they do not like as much.

Genetic diversity

A bee's abilities are influenced by both genetics and the temperature of the brood nest. All bees in a colony are closely related. The queen bee mates with one to two dozen drones, the male bees, from other hives at the beginning of her life, which

means all bees in a colony are either sisters or half-sisters. As male bees come from unfertilized eggs, some of the bees are super sisters, whose genetic material is 75 per cent identical, not just 50 per cent as is the case with 'normal siblings' (including human siblings). As with humans, there is still great genetic diversity among bee siblings, which in turn leads to a variety of preferences.

OPPOSITE: Each bee is inspected after hatching. If it shows signs of disease – such as a crippling virus that prevents the wing developing properly during metamorphosis – it is removed from the hive and discarded.

BELOW: Close-up of a bee's head. The highly mobile antennae and the bee's 'third eye' are clearly visible: three light sensitive simple eyes on top of the head, in between her two compound eyes.

A bee foraging in the wild. Bees visit up to 1,000 flowers a day, which means they visit at least 20,000 flowers over their entire lifetime – and often more than double that.

Spring

The first flight of the winter bees

At the beginning of spring, as the sun becomes stronger, the big moment arrives: the first outing of the year. A bee leaves the hive for the first time after months of confinement. However, the first bees to fly out in the spring are not the new bees, but rather the old ones that have made it through the entire winter, the 'winter bees'. The honey reserves in the hive start to run low towards the end of winter and, in order to produce new reserves, the bees must leave the hive in search of food. Their first flight is also a cleansing flight as it gives them the opportunity to take a bathroom break outside the hive and thus reduce the risk of disease.

A pollen bath. Both plants and bees have benefited from this interaction for millions of years.

The first flight after winter is a life-threatening endeavour for honey bees. Since bees are cold-blooded animals, their body temperature is nearly the same as the ambient temperature. They can only briefly warm their bodies through muscle activity. But if the outside temperature drops, the situation becomes critical for them. Bees can no longer breathe, let alone fly, if their body temperature falls below 10°C (50°F) because their flight and respiratory muscles fail to function below this threshold. This means that a bee on the search for food on a cool spring day is doomed if it gets lost in a shady forest or if the sun disappears behind the clouds for just a few minutes too long.

The leader

Out of the 10,000 to 15,000 bees in a hive to make it through winter, one bee takes the plunge and, on a day that seems just warm enough, decides to fly out into a completely bleak, still wintry landscape in hopes of finding a flower in bloom. This again shows how bees exhibit different character traits. The 'curious' bees and not the 'reserved' ones are the first to leave the hive. While most bees in the hive feel that it is still too cold to risk an outing, a few bees decide that the temperature on one sunny

A forager bee taking flight after a long winter. Bees can fly at speeds of up to 30 km/h (19 m/h).

afternoon is warm enough. These different personalities in the beehive effectively spread the risk for the bee colony as a whole from an evolutionary perspective. If all bees had the same threshold, they might all leave the hive on the same day and, if it turned out to be too risky after all – for example, getting too cold in the afternoon – many might not survive the outing, which would be a severe setback for the bee colony and drastically reduce its chances of survival.

Pollen, an important source of protein

The first bee that dares to venture out, flies out into a cold, barren world, searching for something edible to feed her colony. She searches for flowers in a world that is still on the cusp of spring, and where almost nothing is blooming. In temperate latitudes, there is always a specific plant that kicks off the year. Its blossoms are easily visible, glowing yellow in the treetops: the willow. Willow pollen is one of the most protein-rich pollens that nature has to offer. However, this protein is not intended for

Pollen spreads all over the bee's body. However, the bee can skilfully comb the pollen out of her fur and brush it onto her legs.

the winter bees. Post metamorphosis, they no longer need protein as they are no longer growing. The larvae in the hive, on the other hand, cannot live on honey alone during this early stage of life as it only contains carbohydrates and the larvae need protein to build their muscles. Flower pollen is where all bees, not just honey bees, get their protein. Therefore, it does not matter that willow blossoms contain hardly any nectar. What matters to the bees is the protein-rich pollen, as it is the protein that allows the bees to grow muscle. But the foragers don't collect the pollen for themselves – it's for the larvae and freshly-hatched summer bees in the hive; they need to grow.

Bringing the pollen home

The term 'pollen basket', which is used to describe the structures on the hind legs of bees, is misleading. They are actually combs, not baskets. The bees deftly brush the pollen caught on the hairs of their bodies during flight and scrape it onto the combs on their hind legs. They continue pressing as much pollen as possible onto their legs to transport

The willow is one of the first plants to blossom and is an important source of protein for bees at the beginning of spring.

back to the hive. Studies show that bees even calculate their payload minus the 'fuel', like experienced pilots. This means that, in preparation for their initial reconnaissance flights, bees take in as much honey as possible, storing it in their honey stomachs, or crops, to serve as fuel. However, once they know how far away the flowers they plan to visit are, they ingest only as much honey as they need for the outbound journey. Once they reach their destination, they drink just enough nectar to give them the energy they need for the journey home, thus maximizing the amount of pollen they can transport. Bees can carry a 'payload' of more than half their body weight: 100 milligrammes. By comparison, the most modern cargo aircraft can only transport a 'payload' up to a third of its total weight. Although flying with an additional load is an extremely demanding activity, it is not one bees shy away from – a single bee can visit up to 2,000 flowers every day.

TOP LEFT: Heavily laden with pollen. The packets of pollen often weigh half as much as the bee herself.

BOTTOM LEFT: A wonder of nature. The comb-like structure on the legs, where the pollen sticks.

TOP: Usually, the pollen brought in by returning bees is the same colour because the bees 'tell' each other where to find pollen.

MIDDLE: A hive bee watches the forager bees at work.

RIGHT: A heavily laden forager bee landing at the hive.

Portrait of our main protagonist. Bees remain in constant contact with the comb and even communicate through it – using vibrations.

Newly hatched bees

The first bee of the year – like all other bees that hatch after her – spends approximately the first 21 days of her life exclusively inside the hive. There is plenty for her to do there. Shortly after hatching, she cleans the other cells where eggs are to be laid. A few days later, she is old enough to produce 'sister's milk' to feed the new larvae in the hive. After that, she receives the nectar brought in from outside by the older bees and stores it in cells. Hive bees, which have never left the hive, are constantly busy. When necessary, the oldest hive bees take up positions as guard bees at the entrance of the hive. They check incoming forager bees with their antennae for the hive scent to make sure that no bees from another colony try to gain access to the hive, and they drive away any bees they identify as intruders.

A flexible division of labour

The division of labour among the bees in the hive is very flexible. For example, if no guard bees are needed, then there are not any. If the honeycombs are damaged, many bees pitch in to build new ones. Bees in the beehive mainly do what is necessary, not what an internal clock dictates. If at any given time this means taking over a task intended for another age group, that is not a problem for a bee. But, like all bees, the first bee of the year does primarily what comes most naturally to it.

A bee's first flight

After about 21 days of total darkness, the big day arrives: The first bee of the year to hatch leaves the hive for the first time. Bees spend the first half of their lives in the beehive, only leaving it in the second half of their lifespan. To prepare for this, their bodies undergo another dramatic change. They transform themselves into small yet formidable flying machines. They lose almost all of their body fat and change their digestive tracts so much that their body weight drops from around 100 to just 60 milligrammes. At the same time, they are able to absorb more oxygen to power their

The fine antennal structures contain thousands of sensory cells that are clearly visible.

flight muscles. The amount of glycogen stored in these muscles doubles. This is used as an energy source. Thanks to these changes, they are able to fly back to the hive from over two kilometres away with 20 milligrammes of pollen packed onto their legs or a stomach full of nectar weighing more than half their body weight.

Inside and outside tasks

In spite of the dangers posed by the outside world, bees have no choice but to venture out in search of food for themselves and their colony. The reason they do this at around 21 days of age is due to a risk assessment that has developed in the course of evolution. Every single bee is valuable to the colony as a labourer. In the hive, someone has to build the cells, feed the larvae, and take out the rubbish (bees actually have to do this, too). At the same time, however, bees also have to leave the

safety of the hive to collect pollen and nectar. The tasks inside the beehive are much less dangerous than those outside. As a result, evolution has created an age-based division of labour: bees perform the less dangerous tasks first and then continue with the more dangerous ones. That is why each bee spends the first three weeks of her life in the hive before setting out in search of pollen and nectar. If she does not return from one of her foraging flights, it is obviously not good news for the bee, but the loss to the colony is not as severe, as at least the bee has already contributed to the community for three weeks. Evolution can be cold-hearted at times.

A bee shortly before taking flight. It spends half of its adult life in the hive before finally venturing out into the world.

Predators and other dangers

It is hard to imagine, but bees would not leave the hive at all if they did not have to. The hive provides protection, warmth, food, grooming, and company. Yes, company and possibly companionship. Experiments have shown that a bee, if kept by itself, dies within a day, even when provided with food, water, and warmth. They are made to be part of a collective. In contrast to the safety of the hive, a hostile wilderness awaits them outside, brimming with dangers. The bee-eater is one of these dangers, a colourful bird that specializes in hunting bees in flight, killing them, neutralizing their stinger by rubbing it off on a branch, and then swallowing them whole.

A particularly cruel fate is in store for bees that fall prey to the beewolf, a species of wasp specialized in hunting bees. It lies in wait for bees in flowers and then paralyzes them with its venomous sting. The bee does not die from the sting, but it renders the bee helpless. The beewolf then squeezes out all the nectar the bee has collected and devours it on the spot. It then drags the paralyzed bee into a tunnel, lays an egg on her, and seals the tunnel so its larva can feed on the bee when it hatches.

Perhaps even more astonishing is a species of spider that has perfected ambush hunting on flowers. The crab spider can adapt its body colour to the flower atop which it patiently sits and waits for its prey, with outstretched front legs. When a bee lands within reach, the crab spider quickly snatches her and injects venom into the back of her neck with a bite. Luckily for the bees, not every attack is successful. But bees that have been attacked by a crab spider on a flower and survived remain skittish for life. Experiments related to this predatory method led German neuroscientist Prof. Lars Chittka to suggest that bees might have a more complex psyche than previously assumed. During one experiment, bees that had experienced such an event during a trial abruptly flew away during subsequent trials for no obvious reason, as if they had seen a ghost. Even if the bees did not flee immediately, they approached the flowers extremely cautiously and inspected them from the air for much longer than bees that had never been ambushed by a crab spider.

A bee getting ready to take flight. Her dense fur not only provides warmth but also protects her against water and dirt.

Honey bee senses

Few bees die of old age. Most succumb to the dangers they face in the outside world. To avoid being surprised by bad weather, predators, or failing to find their way home, they need all their senses. At the same time that a bee's body changes into a lightweight, muscular flying machine, its brain also changes. Several days before the first flight, four structures in the bee's brain grow larger. They are called mushroom bodies, and they can grow by up to one-fifth of their volume. The exact functions of these brain regions are not yet fully understood, but we do know that all the information provided by the bee's senses converges and is processed here. It seems to be the control centre of the brain (comparable to the cerebral cortex in humans), where higher brain functions take place. It is also clear that the mushroom bodies are linked to the bee's impressive intellectual achievements. German neuroscientist Prof. Lars Chittka even believes that bees may have something akin to consciousness.

Bees have small scent glands on their feet. In 1992, Prof. Josué Núñez and Prof. Martin Giurfa discovered that they use these glands to leave behind a scent mark on flowers to tell their sisters, 'There's nothing left here.'

A bee collecting pollen from a willow, often the first food source of the year for them. It deftly presses the pollen onto her hind legs while in flight.

Navigation tools in a new world

Before the first bee of the year can embark on her great adventures as a forager, it must first be 'trained'. Winter bees take the newly hatched bees out for a trip around the hive. This is followed by a series of navigation flights, which gradually move farther away. The first thing the bee needs to do is memorize the location of her hive, and the prominent features in the surrounding landscape help with this – objects such as a large tree, a house, or a river near the nest are landmarks to help her find a way home. These first days are crucial and the risks are high; almost half of the bees

old enough to become foragers do not survive this short training phase and never become proper forager bees. Many get lost during reconnaissance flights and never return home. To maintain her orientation, the bee needs to use all of her tiny brain. Navigating the world is so difficult that most of a colony's foraging work is ultimately performed by only 10 to 20 per cent of its bees.

Perceiving the world in slow motion

Bees manage to navigate through their environment using their compound eyes, each of which has 6,000 individual lenses. For comparison, this is like a camera with almost panoramic vision but a resolution of only about 1 per cent of what we have in HD videos. Humans would not be able to recognize anything in their environment with such a low resolution but, then again, humans possess a completely different visual system. Instead of lens eyes with a retina, bees have much smaller and lighter compound eyes. Compared to lens eyes, compound eyes have relatively poor resolution. Bees compensate for this by processing visual information at high speed. Bees see their world, so to speak, in slow motion – a great advantage for a fast-flying insect. In addition, bees only see colours when flying slowly, at speeds below 6 kph (3.75 mph).

ABOVE: View of bees arriving at the hive. Despite thousands of takeoffs and landings per day, collisions are rare.

LEFT: Bees memorize their nest's surroundings very precisely so that they can find it again. They use both large landmarks, such as trees, and small markings at the entrance to the hive.

Altered colour perception

To assess a flower, for instance, bees need to inspect it up close and at their leisure. Bees perceive a slightly different colour spectrum than humans, as bees cannot see red. If bees could look at a rainbow close up, they would not be able to see the red band, but they would see an additional ultraviolet band on the inner arc of the rainbow. Many flowers have adapted their colours to take into account bees' colour perception and display particularly striking patterns that are only visible in the ultraviolet range.

Impressive orientation

In addition to her two large compound eyes, a bee also has three small simple eyes or 'ocelli' on her forehead. These are very primitive eyes that only perceive brightness – the bee cannot focus sharply with them and certainly cannot recognize colours. However, these eyes are invaluable for flight. Thanks to them, a bee always knows where up and down are. The bee's eyes are thus perfectly adapted to its living conditions.

Even though the compound eyes create a less clear image of the surroundings, bees have some tricks up their sleeve that help them recognize things even with these 'weak' eyes. Normally, you need two eyes that are spaced relatively far apart to see in three dimensions. However, the bee employs the same three methods we use to perceive depth even when closing one eye. The first rule is that far-away objects appear smaller than close objects. To use this, the bee must learn how to recognize an object and whether it is far away or nearby. The second rule is that nearby objects move faster than distant ones, and the third rule is that nearby objects obscure distant objects, not vice versa. Based on these methods, bees can orient themselves very well. But their impressive abilities do not end there.

Map of odours

A bee's sense of smell is even more impressive than her sense of sight. Each bee antenna possesses more than 20,000 odour receptors, whose signals reach a region of the brain with 160 small 'glomeruli', which are used exclusively to identify odours. From a neuroscientific standpoint, the sense of smell works the same in all animals, whether they are insects or vertebrates like us humans. A specific pattern of glomeruli – also called olfactory glomeruli – always responds to a specific odour or a mixture of odours. Thus, we all have a map of odours genetically ingrained in us from birth. Humans have 500 glomeruli, yet we cannot identify odours nearly as well as bees. Prof. Randolf Menzel from the Freie University of Berlin tested bees' sense of smell with various chemical alcohols that differed by only one carbon atom, from ethanol

Bees actually have four wings. However, they interlock to form two large wings while flying. With their antennae, bees can perceive the scent of their peers, guiding them to their destination over the last few metres.

with two carbon atoms to hexanol with six and decanol with ten carbon atoms. We humans are not able to distinguish between these odours – for us, they all simply smell like alcohol. However, a bee has no problem differentiating between them.

A bee can also smell in three dimensions, as Prof. Hermann Martin from the University of Würzburg demonstrated. This is something we can only do with our hearing. For example, we perceive that a sound reaches our left ear first and then

Poppies do not offer bees nectar, but they do provide rich pollen. Pollen consists of hundreds of components, and its composition is different for every plant species.

our right ear, concluding that the sound is coming from the left. Almost the same thing happens to bees when they smell something. If two odour sources are next to each other, the odours reach the antennae at different times: first the odour that is closer and then the one that is farther away. The bee can distinguish between them, thus obtaining an accurate odour map of her surroundings. Additionally, she can move her antennae, which have the odour sensors, so as to discover the direction in which the strength of the odour increases or decreases. This allows the bee to orient herself quickly based on the smells in the landscape, easily finding the nectar source on flowers. Back at the hive, it also allows her to determine which of two adjacent larvae is hungrier because the hungrier larva emits more pheromones, signalling that it needs to be fed.

All these sensory impressions allow bees to accomplish one of the most demanding feats of navigation in the animal kingdom. To be successful, a bee must venture into the potentially hostile environment outside of her hive and find small flowers that

A bee on 'capped' honeycombs. The more honey there is in the nest, the more aggressively the bees will defend it.

appear at unpredictable times and are short-lived. After gathering nectar, she needs to find her way back to the nest. This is no easy task. But bees have developed an impressive sense of orientation, which they demonstrate by flying straight home, even after meandering through the landscape in wide arcs and loops while foraging for flowers. Bees always seem to know where home is.

Exceptional intelligence

Bees become smarter as they grow older – they are not wiser or more experienced but actually have more cognitive capacity. As impressive as their senses of sight and smell are, their most important characteristic is their exceptional intelligence. To test this, scientists have let bees fly through bizarre experimental setups, such as labyrinths marked with colours and abstract hieroglyphs. They have disrupted their temporal orientation by putting bees to sleep during broad daylight, only to wake them up later in the day. They have set seemingly impossible tasks for them, such as tracking down sugar water in the middle of a lake even though bees do not fly over large bodies of water. Scientists have also released bees more than 10 kilometres (6 miles) away from their home to find out if they could find their way back from places they had never previously visited. Honey bees have passed most of these tests with flying colours, and every year studies are published that further expand our understanding of bees' cognitive abilities and prove that they are capable of cognitive feats previously attributed only to higher mammals and birds.

Bees, for example, excel at recognizing visual patterns, to the extent that they can distinguish between human faces. This may seem particularly impressive, though for bees, photographs of human faces above a food source are nothing more than strange

Despite the fact that bee vision is very different from human vision, bees can be taught to distinguish human faces and even to distinguish between a Rembrandt and a Monet painting.

flowers. Nevertheless, their ability to recognize complex patterns sets them apart in the animal kingdom. Bees can also recognize similarities among symbols or images and understand them as categories, applying that understanding to completely unfamiliar images. When, for example, some sugar water was placed behind the image of a star-shaped symbol and the star-shaped symbol was then replaced by a completely different star-shaped symbol, bees still recognized that it was star-shaped and assumed that there would also be a reward waiting behind it. Bees can differentiate between artists based on their painting styles. This was demonstrated by Prof. Wu Wen and his team at the University of Queensland, Australia, in an experiment where bees were taught to differentiate between a Picasso and a Monet. They succeeded in no time, even when shown paintings they had never seen before.

Bees' abilities continue to surprise us. When retired chemistry professor Hans Joachim Gross joined Prof. Jürgen Tautz's research group and asked whether bees could count, he was initially met with incredulity. However, the scientists at the Technical University of Darmstadt conducted tests and found out that they can. At least, they can quickly recognize quantities up to four, even if these quantities are represented by completely different symbols. In subsequent work, Prof. Gross found that four seems to be a universal limit. Many organisms and people worldwide have no problem immediately grasping quantities of up to four items; beyond that, they have to look more closely. Perhaps that is why, when tallying marks, we cross the fifth mark through the others, turning it into a visual package to make it easier for us to understand. Bees can also differentiate between 'more' and 'less', regardless of how the quantities are represented. This means they can learn to expect a reward at a symbol board with fewer dots than on a board shown previously. A team from RMIT University Melbourne, Australia, led by Prof. Adrian Dyer and Dr Aurore Avarguès-Weber, even managed to train bees to approach an empty display board after previously showing them a board with only one dot. They seemed to understand that zero dots is less than one dot. This is a skill previously demonstrated only by primates, some birds, and humans.

Bees also have no problem learning rules. Prof. Martin Giurfa and his team at Paul Sabatier University in Toulouse, France, demonstrated that they can apply rules they have learned to completely new situations. When they were supposed to fly towards slanted lines or certain colours, bees did so, proving they can distinguish between similar and dissimilar things. When the scientists then presented the bees with smells instead of symbols, the bees easily applied the rule they had learned to the new situation without further training – 'the reward is behind what is similar to what I perceived at the entrance of the maze' – and chose to fly towards similar smells, where they were rewarded with sugar water.

Exceptional memory

Bees have an exceptionally good memory. Just as the smell of sunscreen can evoke memories of a beach holiday in humans, scents can evoke memories of a particular place and how to get there in bees. They are also extremely good at learning from each other. No honey bee has ever grown up without another honey bee, meaning 'tradition' – the transmission of skills outside of genetics – plays a significant role for bees. A bee comes into the world as a blank slate. It must gradually learn everything that is important for its own survival and that of the colony, such as what flowers look like. As with us, there are also different types of problem solvers among bees. Prof. Chittka and his team showed that some bees are consistently quick and sloppy, while others are more cautious, slower, and more accurate in their decision-making. Both strategies are justified depending on the challenges posed by the environment. Sometimes the fast, sloppy decision-makers have the advantage (for example, when quickly exploiting suddenly emerging food sources), and sometimes the slow and careful ones (for example, when avoiding ambush predators). As a whole, a bee colony performs best when it harbours a variety of individuals who possess as many different strategies as possible. When conducting experiments on bee learning behaviour, there are often one or two 'geniuses' who solve a problem faster than all the others, or in an exceptionally efficient way, or in a way completely unexpected by those organizing the experiments. In one experiment, where Prof. Chittka was examining the efficiency of bee foraging in the wild, each bee was weighed when leaving the nest and again upon its return to assess how much nectar it had collected based on the weight difference. To do this, each bee had to be briefly captured in a black plastic container when leaving the nest and again when it returned. Most bees showed some aversion to being captured, but some soon grew accustomed to the procedure. However, one bee regularly flew directly into the black container, even if the experimenter held it upside down a few metres away from the beehive. Presumably the bee simply viewed the container as 'public transport' and expected to be carried back to the nest inside it.

Are bees conscious?

This question is difficult to answer. The neuroscientist Randolf Menzel prefers to use the term 'intentionality'. According to him, bees do have expectations and make decisions based on those expectations. One piece of evidence for this is an experiment conducted by Menzel and his team where they offered bees a bowl of sugar water near their hive. After the initial forager bees found and inspected the food source, the scientists replaced the sugar water with a much less sweet solution. Shortly afterwards, the first bees appeared at the feeding site, having learned about it only from their fellow bees. These bees were hesitant about taking the sugar water,

Thousands of takeoffs and landings per day – foreign bees are recognized by a pheromone signature.

if they did so at all, and searched the immediate surroundings for alternatives. They were, so to speak, disappointed by the sugar solution as they had received something much sweeter from their sisters. Their expectation was not met, so they continued looking for what they were expecting.

Being a slower learner is beneficial

Being highly intelligent, and having the associated ability to learn rapidly, seems to have so many advantages and hardly any risks. Thus, one might ask – at least as a biologist – why are there slower-learning bees at all? Why was this not eliminated by evolution long ago? Prof. Nigel Raine from the University of Guelph in Canada has the following answer regarding bumble bees. In a study, faster-learning bumble bees were active on fewer days than slower learners. The difference accumulated to the point where the less intelligent individuals actually contributed more in foraging terms than the smarter bees, and thus more to the colony over the course of their lives. So, without the slower-learning individuals, the bee colony would be worse off.

Learned and new behaviours

The ability of bees to learn and adapt explains their success in evolutionary terms. Completely rigid, hard-wired, predictable behaviour is a sure path to extinction. If an animal behaves completely predictably when facing a predator, for example, the predator will quickly adapt its own behaviour accordingly and soon be very successful. To survive, behaviour must always remain somewhat unpredictable and give an animal the opportunity to cope with new situations or behave differently in known situations.

A bee's brain is about the size of a grass seed, and yet it contains nearly a million brain cells (for comparison, the human brain has around 90 billion brain cells). While modern computer chips have over three billion transistors etched onto silicon chips, with a few nanometres of spacing between transistors, the transistors have far fewer connections to each other than the neurons of a bee brain because each neuron is connected to thousands of others. The level of complexity of a bee brain thus far surpasses that of a computer chip. Additionally, a bee's brain consumes only one twenty-thousandth of the energy of a computer chip. This means that bees' brains are far more complex and more efficient than our computers.

Protostomes vs deuterostomes

When studying the intelligence of bees, it is worth noting that bees represent a fundamentally different category of animal to humans – and this does not just refer to the differences between insects and mammals. The difference is much more fundamental and is the reason why a bee's brain is structured completely differently than that of a human. Since prehistoric times, the animal kingdom has been divided into two basic categories: deuterostomes and protostomes. The former have a spinal cord: their central nervous system runs along their back. The latter, the protostomes, have a ventral nerve cord: their central nervous system runs along the underside of their body. These two evolutionary lines diverged about 680 million years ago – until then, there were only animals that had a 'top' and a 'bottom' like today's jellyfish, but no 'right' or 'left'. It was not until after this divergence that animals emerged with symmetry on both sides of their bodies, meaning their body structure also had a 'right side' and a 'left side'. However, evolution produced two solutions for this type of symmetry: animals were, anatomically speaking, either on their backs or on their bellies. Deuterostomes, whose central nervous system lies on the back, include all vertebrates: birds, fish, reptiles, amphibians, mammals, and us humans as well. Protostomes, whose central nervous system runs along the belly, include worms, snails, octopus, crabs, and spiders, as well as insects like bees. The intelligence of bees, which may seem somewhat familiar to us – perhaps because of its social component – is actually the result of an evolutionary path completely separate from our own, spanning hundreds of millions of years. Except for octopus, we currently know of no other representative of the other evolutionary line whose intelligence approaches that of bees. Therefore, for neuroscientist Randolf Menzel, bees are at the forefront of their evolutionary line, while we humans, with our cognitive abilities, stand at the forefront of the other line, looking from one peak of evolution to the other, so to speak.

The waggle dance

The next task for the first newly hatched bee of the year is to search for food. It does not have to do this alone. Forager bees who have found an attractive food source advertise their discovery and guide newcomers to the location using a series of gestures and signals. It all starts in the darkness of the hive with what are referred to as waggle dances. Returning forager bees run in a tight circle or in a figure-eight pattern atop a small dance floor on the comb. Along a short straight line, they rapidly waggle their bodies conspicuously back and forth. As the fine vibrations spread across the comb, other bees are alerted and follow the dances if they are in the mood to forage. The follower bees sense and feel the movements of the dancers with their antennae. Biologist Karl von Frisch recognized that the movement pattern of the waggle dances provides clues as to the location of the target outside. He also realized that bees use a reference point for orientation, one that humans have been using for navigation since ancient times: the sun. The dancing bee indicates the angle at which the follower bees should fly in relation to the sun. However, because it is dark in the hive, bees have to come up with a reference point to symbolize the missing sun.

The solution is as simple as it is ingenious: the direction opposite gravity, or 'up', symbolizes the direction to the sun. If the bees should fly towards the sun, the lead dancer waggles vertically upwards on the comb; if they should fly away from the sun, she waggles vertically downwards. The same applies for other directions. The distance to a target is also represented by the duration of the waggle run. This communication is only possible because honey bees build vertically hanging combs. Without its vertical combs, the honey bee could not use gravity to represent the

position of the sun and thus communicate the angle at which a bee should travel in relation to the sun in order to locate her destination. Bees that do not construct such combs, such as bumble bees and tropical stingless bees, have not been able to develop the unique waggle dance of honey bees.

The bee odometer

Karl von Frisch received the Nobel Prize for his work in 1973, and in the 50 years since, detailed studies of bee communication have shown how complex the process actually is. The dance conveys only rough information about the location. The angle and distance indicated for the same target vary greatly and cannot be used as the sole

The bee dance is the only known symbolic language in the animal kingdom. However, according to recent findings, it is only part of a series of information that bees share with each other.

means for finding the destination. A bee's impression of the distance flown, which is included as distance information in the dance, changes depending on the landscape. The 'bee odometer' is based on what is called optical flow: the images seen passing in front of the bee's eyes. We humans can estimate fairly accurately the speed of travel from the optical flow that occurs when staring out of the window of a moving train, but not the distance travelled. Bees can do this, too, albeit very roughly. If the landscape is monotonous, the distance appears short; if it is varied, it appears long. These distance indications are different for each bee subpopulation. This was determined by Patrick Kohl and Benjamin Rutschmann, biology doctoral students at the University of Würzburg, in an experiment where they observed different bee species in the same experimental setup. They offered sugar water at two different distances, one after the other: first relatively close to the beehive and then four times farther away. For the eastern honey bee, the waggle dance lasted half a second for a target 100 m (330 ft) away and two seconds for a target 400 m (1,300 ft) away. However, for the giant honey bee, the waggle dance for a target 100 m (328 ft) away also lasted half a second, but when the feeding site was 400 m (1,300 ft) away, the waggle dance was just one second.

Geographical peculiarities

There are also differences between the different populations of western honey bees in Europe and America, as well as between the nine honey bee species. Bees south of the Alps indicate distances differently than bees north of the Alps. Thus, even among honey bees, there are dialects. Nevertheless, the symbolic language of bees still works well; even if the angle indicated by the dance is not precise, and its duration only indicates the approximate distance, additional signals can help the recruited forager bees reach their destination in the field.

The apple tree – object of desire

Another helpful aid is the scent of the advertised flowers, since it sticks to the dancer until it returns to the beehive where other bees perceived it. The dancer also provides additional help. Once she reaches the destination, she opens a gland on her abdomen, which emits a scent that can be detected by other bees searching for the destination. This leads the forager bees to the destination like a beacon. Thanks to this communication, honey bees are extremely efficient at harvesting nectar and pollen – and thus at pollination. While nectar and pollen collectors like butterflies, bumble bees, and other wild bees have to laboriously find each flower on their own and more or less at random, honey bees inform each other that a tree is in bloom. Indeed, their effectiveness in locating and exploiting such food sources is unparalleled. It is a well-known fact among beekeepers that honey bees can, as a swarm, 'strip bare' a blossoming apple tree within a few hours.

An apple tree in full bloom is a rich source of food, prompting the first forager bees to report their find back to their sisters.

The origin of bees

Science continuously grants us new insights into the origin of bees. Broadly speaking, they evolved approximately 120 million years ago from wasps. Wasps met their protein needs by hunting and preying on small insects called thrips (and they are still predators to this day). When dinosaurs roamed Earth, there were only a few plant species that produced pollen. Plants with pronounced flowers, such as roses, orchids, and fruit trees, did not yet exist. Mosses, ferns, and conifers dominated the world, some of which produce seeds but were pollinated by the wind – a very risky and inefficient strategy from the plants' perspective. They had to expend enormous amounts of energy to produce large quantities of pollen or spores, only for the wind not to blow or for it to rain. From an evolutionary perspective, it was a very 'costly' strategy.

Flowering plants, on the other hand, were much more efficient. They only had to produce a small amount of pollen for a relatively short time, which attracted flies, butterflies, or beetles to do the pollinating. After a short period of time (in evolutionary terms), specialized pollination service providers arrived, collecting the pollen packets and taking them to the next plant – an ideal solution. Some wasps eventually covered their protein needs entirely by consuming these pollen packets – thus, bees were born. About 120 million years ago, the first solitary bees appeared, together with the first primitive flowers. This marked the beginning of a dramatic development that continues to shape our planet to this day. With the later appearance of social honey bees and other pollinating insects, the evolutionary 'explosion' of flowers started. Even the asteroid that wiped out the dinosaurs 66 million years ago could not stop the spread of flowering plants. Today, there are over 120,000 species of flowering plants – including the most successful plant families ever – and almost all of them depend on insects for pollination.

In contrast, there are currently only a few hundred species of non-flowering plants in existence, as they simply cannot compete with flowering plants. Humans depend on these plants for a large part of food production. While wheat, rice, and maize – the most important food plants – belong to the family of grasses and are thus self-pollinating or are pollinated by the wind, most fruit, some vegetables, bell peppers, sunflowers, and many other food plants depend on insects for pollination. Even if insects are not strictly necessary for reproductive purposes, they often increase yields dramatically or are necessary for the production of seeds.

A matter of sweetness

Back in the hive, nurse bees – younger bees that are not yet foragers – take the sweet load delivered by the new forager bees and process it. Bees often feed each other and pass the contents of their crop to other members of the colony. How long a bee has to wait for nectar to be taken after returning to the nest is crucial, as this influences the forager bee's motivation to fly out again and collect more nectar. The longer she has to wait, the less motivated the bee is to gather more of what it has just brought in. Thus, the balance of supply and demand in the nest automatically regulates the provision of the necessary resources. As Prof. Tugrul Giray of the University of Puerto Rico has found, there is not only a genetic difference between bees that collect pollen and those that collect nectar, but there are also further differences among the nectar gatherers, with some preferring sweeter nectar than others. Each bee has its own preferred level of sweetness.

There are even some individuals that search for food outside but never quite find the sweetness they prefer and thus return to the hive with an empty crop. If they return several times from foraging trips without success, they may even stop foraging entirely. However, this behaviour is only possible when honey reserves are plentiful in

A bee's central mouthpart contains her tongue and proboscis. It remains pressed against the body during flight.

For a long time, bees were thought to be solitary creatures as soon as they leave the hive. But we now know that bees are social butterflies outside the hive as well.

the hive. If reserves are low, the lazier foragers are urged to go out and collect nectar again by their fellow bees vigorously shaking their abdomens (what scientists call 'dorso-ventral abdominal vibration' or DVAV). If they return multiple times without nectar, the continued abdominal vibrations of their fellow bees may also lead to them being less selective about the type of nectar they are prepared to collect.

Few dare to fly through the rain

Bees are keenly aware of the ambient temperature. When the temperature suddenly drops and rain is imminent, bees quickly make their way back to the hive. Raindrops can be as heavy as half the weight of the bee herself. And a direct hit from a raindrop – be it in flight or on a flower – is very dangerous for a bee. The impact itself can stun a bee, and the weight of the water can make her so heavy that she can hardly fly. But most importantly, a wet bee has difficulty breathing. Like all insects, bees breathe through small openings in their exoskeletons called spiracles. Water can block these openings, causing the bee to suffocate, or rather drown.

Experienced bees can sense changes in the wind that predict bad weather.

However, despite these risks, there are some bees in every hive that actually like to fly around when it is raining. Remarkably, genetic quirks can result in a few individuals in every bee colony that are specialized in venturing outside in the rain, even if the rest of the colony wisely remains in the safety of the hive. This individual behavioural diversity among bees within a hive provides an evolutionary advantage: if all bees were to venture out only in the same conditions, new food sources might never be discovered. The variance in the bees' foraging preferences (travelling great distances in search of food or staying close to the hive, leaving the nest in the morning or in the evening, flying in sunny conditions or in rainy weather, etc.) helps the colony adapt to changing vegetation due to changes in the climate. If the food supply varies because the availability of flowers changes, for example, due to a cold spell or an unusually hot period, there is still a chance that at least some members of the colony will be able to locate new food sources because they have 'unusual' foraging patterns. In a sense, bees have 'understood' that it is beneficial in the long run to have individuals within a colony that follow unusual or seemingly unsuccessful strategies, just in case the previously successful strategies of the majority cease to be effective.

A young and inexperienced bee may get caught by surprise in a rain shower.

If a bee is caught outside in the rain, she has no choice but to seek shelter under leaves or an overhang; this sometimes happens to young, inexperienced bees. However, rain clouds do not prevent a bee from finding her way home because, even though she uses the sun to navigate, the sky does not have to be blue for the sun to be of use. Earth's atmosphere scatters and filters sunlight depending on the angle at which the sun's rays hit the atmosphere. A bee's compound eyes are capable of detecting polarized light, so it can perceive the patterns created in the sky and therefore knows where the sun is, even if it is hidden behind clouds. Only after sunset can bees no longer make it back to the hive. Then, the bee has to spend the night outdoors and hope that the temperature does not drop below 10°C (50°F) so that she can return home the next day. Any colder and she will not survive the night.

There are also rain specialists among bees.

Spending the night outside the hive

Some bees even spend the night outside the hive intentionally. Scientists at San Francisco State University observed how some bees did not return to the hive at night but instead huddled under a leaf. This behaviour had previously only been observed in bumble bees, which are aware when they are sick and therefore stay away from the colony to prevent the disease spreading among their fellow bees. In this particular case, however, the observations of the Californian researchers were even more surprising. After a few weeks, the cause for the voluntary isolation became clear to the team. After the bee died, the larvae of the zombie fly (*Apocephalus borealis*) emerged from her body. This widely distributed insect measures only a few millimetres in length and is classified as a parasitoid among dipterans. The fly has a characteristic bump on its head. The adult zombie fly ambushes bees on flowers

TOP LEFT: Due to the low temperatures at night in temperate regions, a night outside the hive is usually a death sentence for a honey bee. In warmer regions, it is possible for a bee to survive a night outside the nest.

MIDDLE LEFT: Bees are immobilized by the cold.

BOTTOM LEFT: A few rays of sun are enough to raise a bee's body temperature, enabling her to fly.

ABOVE: Bees have no problem readjusting to the fact that the sun is now in the east when finding their way home.

and needs only a few seconds to lay its eggs inside the bee's respiratory openings. The parasite's offspring then have time to develop and grow inside the bee's body. In the final phase, the fly emerges from the body of its host – a brutal but natural cycle.

Even bees have to sleep

Back in the hive, the bee, like all animals, goes to sleep. The discovery that bees sleep, indeed that insects in general sleep, was made in the early 1970s by Prof. Walter Kaiser from the Technical University of Darmstadt. It is now clear that sleep is a basic need for all organisms, but it still puzzles scientists. The first observations of sleeping bees were made much earlier. Pliny the Elder documented this in the first century AD but was not taken seriously by scientists for a long time. Why shut down the body, slow down responses, and disconnect the senses? For most animals, this is a highly vulnerable state to be in. Science still cannot fully answer this question. Studying sleep has proven challenging because different species exhibit a wide variety of sleep patterns. Many birds (and also marine mammals like dolphins or seals) sleep with only one hemisphere of the brain at a time, switching between brain halves during the course of the night. But even 'normal' sleepers that, like humans, sleep with both hemispheres of the brain simultaneously display mysterious behaviours. All animals seem to have deep sleep stages and rapid eye movement (REM) sleep stages, during which the eyes move rapidly behind the eyelids. At least in humans, this is the phase in which they dream. Bees cannot move their compound eyes, but they have a rapid antennal movement (RAM) phase, sometimes moving their antennae in exactly the same way as they did when visiting a particular flower during the day (and this varies, with different movement patterns for different flowers). It is hard not to conclude that bees are dreaming of visiting flowers while sleeping. But of course, we will never know for sure. At some point, however, bees enter deep sleep and their antennae stop moving completely.

Bees wiggle their antennae while they sleep like when visiting a flower. It is thought that bees dream about flowers.

Bees sleep more than eight hours a day on average; young bees in particular are late risers. When they are ready to sleep, they find an empty cell and crawl inside.

Young animals are long sleepers

Bees sleep a lot. Young bees sleep for nine to ten hours per day and older ones for six to seven hours. Young animals of all species sleep significantly more than their adult counterparts. Bees also choose different places in the hive to sleep, depending on their age. Young bees like to sleep in empty cells, while older bees tend to sleep at the edge of the combs. Bees probably sleep for the same reason as other animals – so that their brains can process the day's experiences. If they are disturbed during sleep, they cannot process the experience in a way that allows them to efficiently retrieve the information at a later time – in other words, they cannot learn well. Even sea snails, whose brains have only 20,000 neurons (a fifth of what honey bees have), can learn, but they need to sleep in order to process their experiences and learn from them. Bees, with their powerful brains and impressive cognitive abilities, need sleep even more. Their dancing in particular suffers if they do not get enough shut-eye. Studies by Prof. Barrett Klein at the University of Wisconsin have shown that bees that do not get enough sleep are clumsy dancers. Bees have even been observed to notice the inaccuracies in the dances of tired bees and turn away from them to watch another bee that is dancing more adroitly.

The force of a raindrop can sometimes stun a bee.

Flights cancelled due to bad weather

Occasional rainy days are no problem for a bee colony. Once bees have learned that there is a food source available at a certain place and time – such as flowers that have blossomed and are full of pollen or nectar – they will return to them to check if they still have food to offer. After all, it is much easier to visit established food locations than to discover new ones. Even if a food source has dried up because the flowers have withered, or the plants have been mowed down or eaten, they will continue to check that location for a couple of days to see if the food supply returns. If it rains, the period during which the bees will check again is extended accordingly. If it rains for two days and the bees cannot leave the hive, they will still check the feeding site for possible flowers on the third and fourth days. If it continues to rain, bees can even wait for up to a week to visit the feeding site again when the sun is shining and see if there is anything left to collect. The bees' sense of time is impressive. Prof. Martin Lindauer from the University of Würzburg demonstrated that bees that are regularly offered sugar water at 6 p.m. remember this time and search for the reward at 6 p.m.

at the same location. If the reward is offered to another colony at 8 a.m. at a specific location, the bees will be waiting at that location at 8 a.m. sharp. Even the larvae of the beehive that was accustomed to feeding at 6 p.m. will also go out in search of food at 6 p.m. once they become bees, while the larvae from the other hive will do so at 8 a.m. once they have completed their metamorphosis.

Prof. Tugrul Giray was able to demonstrate that honey bees are among the few animals that can be active both during the day and in the morning and evening. A bee colony essentially organizes its workday in shifts, with some bees rising early and others only becoming active for the late shift. This is a preference the bees inherit through the paternal line, like their preference for a certain degree of sweetness or their impulse to heat the hive at a specific temperature.

Sharing

If a bad spell of weather persists for an extended period of time, it can pose a serious challenge for the bees. Rain prevents *most* bees from foraging, but even sudden drops in temperature or strong winds can ground the bees by making it too difficult for them to forage. The later in spring a longer period of bad weather occurs, the more dangerous it is for the colony since this is the time when the colony is growing

A honey bee dives into a honey cell to find honey at the bottom of the cell. Even if the honey supplies are low, all the honey is shared fairly within the bee colony.

TOP: Honey bees redistributing honey within the colony. Honey bees regurgitate honey and share it with bees from the same colony. In this way all the honey reserves are shared evenly among all members of the colony.

ABOVE: Once the honey reserves have been exhausted, all the bees in the colony starve to death within a few hours.

and thus is most in need of food. If the forager bees cannot venture outside because of the rain, the bees have to feed exclusively on the honey stored in the hive. And these supplies can dry up quickly. When that happens, honey bees exhibit a logical but still impressive behaviour: they share all supplies equally among all members of the colony, down to the last drop of honey. Survival is not reserved for only the strongest or a select few. Instead, all bees in the colony receive equal shares of the honey supply. Once the honey has been exhausted, the entire colony perishes within a relatively short period of time – sometimes in just a few hours.

The challenge of climate change

Climate change poses a significant challenge for bees. Short-term extreme temperatures are their least concern, as are changes in average temperatures. The primary danger lies in shifting seasonal weather patterns and disruptions to nature's rhythms. Extended periods of cold or rain late in spring not only hinder foraging but can also damage flowers and alter vegetation cycles. These shifts are particularly dangerous for non-social wild bee species, as they are far more dependent on external environmental cues than honey bees. The latter benefit from the care of beekeepers, who can partially mitigate these effects. That being said, these challenges add to the growing list of stress factors for honey bees, which are already struggling with parasites, food scarcity, and exposure to environmental toxins.

Social bonds

Bees rarely fly alone; they often prefer to venture out in small groups. These small 'gangs' typically consist of three to five bees, and the individual gang members tend to remain the same, especially when visiting the same feeding sites. Prof. Tugrul Giray has documented that these social bonds remain even when the group moves on to explore new food sources. While the term 'friendship' is seldom used in scientific contexts as it risks anthropomorphizing animal behaviour, it is clear that bees do prefer to spend time with certain individuals more than others. Prof. Mehmet Ali Döke from Utrecht University found that bees actively support each other in difficult situations, whether they are injured, hungry, or facing another challenge. Remarkably, this altruistic behaviour is not limited to life inside the hive. It also occurs in the wild while bees are foraging for food. For instance, a bee nearing exhaustion may receive a small share of nectar from another member of her colony, providing just enough energy to make the journey back home. And if a bee encounters a trapped hive-mate, she will often attempt to free it. Similarly, thirsty bees may receive water from other bees. It is surprising that science viewed foraging bees as solitary creatures for so long, despite the well-documented social nature of honey bees within the hive. Recent technological advances, such as radar tracking and radio frequency identification (RFID) tagging (a technology we know from 'chipped' pets), have allowed scientists to observe individual bees more closely in their natural environment. These tools have revealed that, even while foraging, bees recognize, communicate with, and support other members of their colony. This new understanding highlights the complexity of bee social behaviour, both inside and outside the hive, offering fresh insights into the cooperative nature of these remarkable insects.

How far do bees fly?

To locate new food sources, preferably large clusters of flowers such as blooming trees or expansive fields, the hive relies on scout bees – individuals that fly farther than most of the others. The likelihood of a bee becoming a scout appears to be influenced by her genetic makeup, notably linked to the lineage of the drone from which she descends. The typical foraging range of a bee colony extends up to 3 to 4 kilometres (2 to 2½ miles) in radius. No bee would fly beyond the capacity of her energy reserves. In other words, each bee is acutely aware of the distance she has flown from the hive and the amount of nectar she carries in her crop, which directly correlates to her remaining energy.

However, within every bee colony, there are individuals that push the limits of this flight radius as much as possible. While most bees collect pollen and nectar within a one-kilometre radius, they are capable of flying up to 3 kilometres (2 miles) or more if necessary. If no flowers are found within their immediate range, some bees will fly as far as 5 kilometres (3 miles) to find some. Each bee has her own unique foraging strategy: Some search intensively in the immediate vicinity, while others fly directly to more distant areas. The most daring scouts venture even farther, pushing beyond the normal limits. These long-distance scouts can only sustain such extended flights by making small 'refuelling stops' at individual flowers along the way. These stops provide them with just enough nectar to continue their search but are typically not significant enough to warrant a return to the hive to report their findings.

Only about 5 per cent of the bees in a hive ever discover their inner scout by venturing out alone to search for new food sources. Research by Gene Robinson from the University of Illinois revealed that when bees successfully find new flowers, the same genetic pathways and chemical compounds are activated in their brains as in mammals when experiencing something exciting and new: so-called thrill-seekers.

In extreme cases when there is no food nearby, bees will forage up to 10 kilometres (6 miles) but never further.

In the same way that humans experience a rush of adrenaline while bungee jumping, scout bees experience a rush of dopamine after locating a blooming apple tree.

In a separate study, Prof. Lars Chittka and his team tracked the foraging behaviour of individual bees by attaching small radar reflectors to their backs. They followed the bees starting from their maiden flights, through their daily foraging trips, and ultimately to their deaths. The researchers observed distinct differences in their foraging strategies. Some bees foraged in only two locations, after conducting just two initial exploratory flights, over the course of their entire lives. In contrast, other bees never settled on a single feeding site, with almost every foraging flight being an exploratory mission even though well-known feeding sites full of flowers were available and regularly visited by other bees. These scout bees are fearless explorers, willing to take risks in search of new resources. Their daring behaviour carries its risks but can also be vital to a colony's survival. By discovering especially abundant food sources, these scouts can significantly improve the chances of survival for their entire colony.

Although lavender is bee-friendly, it is often contaminated by pesticides.

Pollen is not just pollen

The reason bees search so extensively is not just to find flowers. Bees need a wide variety of pollen because pollen is not just pollen. Each plant species contains a large number of very different substances and this holds true for their pollen as well. The pollen of each plant species has very a different composition – pollen of a daisy is not the same as that of a sage. By weight, some pollens are made up of one fifth sugars and half protein, in addition to minerals, trace elements, almost all the vitamins, and lots of different healthy fats. The co-evolution of pollinators and flowering plants has ensured that plants have made their pollen into the perfect food for bees. Experiments conducted by a research group led by Prof. Jürgen Tautz have shown that the health of a bee colony significantly decreases when bees have a monotrophic diet. Resistance to disease and parasites decreases, and the overall condition of the colony suffers. Therefore, when there is a sufficient variety of flowers available, millions of years of evolution have enabled bees to take advantage of this unique food source. This makes it is all the sadder that the increasing destruction of rural vegetation in industrialized nations has led to a lack of food for bees.

Monocultures in agriculture – such as rapeseed – bloom for only a short time, and the supply of wild, diverse vegetation that blooms at different times of the year has dramatically decreased, especially in rural areas. Beekeepers a century ago could not have imagined that it would come to this, but bees are already starving in the middle of summer because there are no longer enough flowers nearby to feed them. Bees have proven in their evolutionary history that they can adapt very well to a changing environment. Nevertheless, climate change, environmental toxins, parasites, and, above all, a lack of available food sources are currently making life very hard for them. Every living thing prefers certain living conditions. This can mean a specific temperature, a special food source, or the absence of certain predators. Accordingly, there are also conditions under which survival becomes increasingly difficult. A certain environmental toxin or lack of food may not be fatal straightaway, but it can certainly make life challenging. When many of these negative factors come together, more and more animals – or in our case, colonies – die. Bees, like all insect species, are adapted to certain environmental conditions in terms of their developmental stages and behaviour. If these conditions change, the most sensitive ones die first, then the less sensitive ones, and so on. The pressure simply becomes too great. Perhaps the hardiest bees will adapt and survive, but the conditions humans are currently creating for bees and many other insects are pushing many species to the brink of extinction and beyond. If one or more bee species becomes extinct in a region, a gap is created in the ecosystem, and other species that are naturally adapted or can adapt quickly will move in.

Due to changes in Earth's climate, the pesticides being introduced into the environment, and the disappearance of many flowering plants, humans are decimating the insect world and drastically reducing its diversity. It is estimated that one-third of all insects worldwide are threatened with extinction. Pollinators are particularly vulnerable, and 75 per cent of our food crops depend on these pollinators. However, insects have an extraordinary arsenal of biochemical compounds, many of which are still unknown to us, which they use to defend themselves against viruses, bacteria, and fungi, helping them to survive. Indeed, in the fight against highly resistant bacterial strains and other pathogens, these substances could still be of great value to us, and this treasure trove has yet to be fully exploited. But these interesting compounds are not the real loss. Many ecosystems humans need to survive rely on insects. Their disappearance would have dramatic consequences. If insects die out, then the entire network of closely intertwined animals and plants is in jeopardy.

LEFT: Meadows that are mowed too early pose a problem for bees. Especially in rural areas, there are often not enough flowers to feed the bees.

TOP RIGHT: Rapeseed is beneficial for bees but also problematic because a rapeseed field only blooms for a short time. After that, there is no longer any food available for the bees.

MIDDLE RIGHT: Cornflowers provide bees with pollen and nectar.

BOTTOM RIGHT: Honey bees sometimes observe bumble bees using their strong jaws to bite a hole into the base of a flower to access the nectar. Once the bumble bees leave, honey bees feed on the leftover nectar that they would not have been able to reach without the help of the bumble bees.

OVERLEAF: Just as every plant is made up of different components, the pollen of each plant also contains different nutrients. A wide variety of pollen is important for the health of a bee colony.

The death of the winter bees

By late April or early May, the bees that endured the winter finally reach the end of their lives. Having spent months heating the hive to ensure the queen's survival, they finally die sometime in spring. These winter bees typically live for around six months – far longer than their spring-born counterparts, who survive for only

When a bee discovers a dead companion in the hive, she grabs her with her mandibles… and swiftly carries her outside to dispose of her body.

Certain bees remove dead bees from the hive more frequently than others. These individuals are aptly referred to as 'undertakers'.

TOP: After about six months, the bees that were born in the previous autumn reach the end of their lives. Their lifespan is nearly four times longer than that of summer bees.

ABOVE: Bee wings endure significant wear and tear from continuous flying. The turbulence generated at the wingtips places considerable strain on the delicate wing structures, gradually weakening the material over time.

about six weeks. While few bees die inside the hive, the rare occurrence of such deaths poses a significant risk. Decomposing bodies can foster the spread of bacteria and fungi, threatening the colony's health. To mitigate this danger, bees diligently remove deceased hive-mates, just as they meticulously dispose of faeces and other potentially harmful residues.

Are bees alike?

This raises an intriguing question: can bees recognize individual members of their hive? Simply put, we do not know. No study has tested this yet. However, we do know that bees possess an extraordinary capacity for recognizing patterns. For example, they can identify photographs of human faces – an ability thought to stem from their remarkable talent for recognizing flowers. But this does not necessarily mean that they can identify individuals within their colony. We also know that paper wasps of a certain species can tell apart individuals within their group. Research by Prof. Elizabeth Tibbetts at the University of Michigan, USA, showed that these wasps could recognize each other, even after a week apart. This social species, which lives in much smaller groups than bees, relies on facial recognition rather than scent to identify colony members. Furthermore, the wasps were able to recall whether previous encounters with a specific individual were positive or negative. Whether or not bees share this ability remains unknown. What we do know is that bees can differentiate between their full and half-sisters.

Pain and other emotions

Another question is whether bees feel emotions. Let us start with pain. The renowned bee scientist Karl von Frisch believed that bees could not feel pain. His conclusion was based on a rather harsh experiment: Frisch used scissors to sever the abdomens of bees feeding on a sugar solution. Surprisingly, the bees showed no apparent distress. Instead, they continued to drink the sugar solution, which now leaked from the severed ends of their intestines, forming small puddles behind them. Based on their apparent indifference to such a severe injury, Frisch concluded that bees do not feel pain.

Modern neuroscience, however, offers a more nuanced perspective. Pain perception is thought to be based on eight criteria of increasing complexity. First, specialized sensory cells, called nociceptors, must detect pain. Second and third, nerve pathways and a specific region in the brain must process the sensation. Fourth, the animal must respond to painkillers, including natural ones produced by its body. Fifth, it must exhibit behaviours that respond to or avoid pain. Sixth, it must demonstrate the ability to learn from pain. Seventh, it must be able to modify the pain response, for example, by increasing or decreasing the reaction depending on the context (e.g. distraction or stress). And lastly, there must be an emotional response to pain. While this is listed as the eighth criterion, it seems more like the result of the other seven. Because emotions are a subjective experience, they are virtually impossible to measure, but the other seven neural and behavioural criteria suggest an emotional experience. Of these eight (or seven) criteria, bees fulfil at least four (except in their abdomen).

Science has paid limited attention to insect pain perception, but this is gradually changing. Bee researcher Prof. Matilda Gibbons, from Queen Mary University, UK, developed an experiment that suggests bees exhibit more than simple reflexes when exposed to painful stimuli. She presented bumble bees with a sugar solution at a feeding station where the landing surface was heated to 55°C (131°F) – a temperature likely perceived as painful by bumble bees because they usually avoid it. Remarkably, the bumble bees chose to land on the hot surface despite the discomfort, motivated by the reward of sugar. This behaviour indicates that bumble bees may consciously evaluate pain stimuli and adjust their actions accordingly. In essence, they are willing to endure 'hot feet' for a taste of sugar water.

Insects are not yet protected under animal welfare legislation, which mandates the protection of animals against 'unnecessary pain'. But this could change. Not long ago, mice were not protected within the framework of animal welfare either, and even with humans the understanding of who experiences pain was remarkably limited. As recently as 1988, a survey among anaesthesiologists revealed that 90 per cent rarely or never administered opioid painkillers during surgeries on newborns – a tragic underestimation of pain sensitivity, even among medical professionals.

Establishing that an organism 'feels' pain, physiologically and emotionally, is challenging. But Melissa Bateson and Jeri Wright, along with their team from Newcastle University, UK, delved even deeper into exploring the emotional world of bees. They employed a classic method from experimental psychology: the cognitive bias test. This test investigates whether a negative emotional state, such as fear, influences decision-making – essentially assessing whether an individual views the world more pessimistically when distressed.

In their experiment, the researchers trained bees to distinguish between two odours. An alcohol-based scent was associated with a reward of sugar water, while a flowery scent (scientifically known as 2-octanone) led to a bitter liquid, which the bees avoided. When exposed to these scents, the bees responded by either extending their proboscis in anticipation of the sugar reward or recoiling to avoid the bitter substance. The scents were delivered via a small bulb syringe.

To test the impact of stress on emotional decision-making, half of the bees underwent a brief but unsettling experience: they were shaken in a small container for 60 seconds. Shaking is known to be uncomfortable for bees, mimicking what occurs when a predator, such as a badger or bear, disturbs their nest. This may serve as a warning signal in the wild, and it could also induce a state similar to fear. The researchers did not rely on assumptions alone. They analyzed the shaken bees' hormone levels and found significantly reduced levels of neurotransmitters such as octopamine, dopamine, and serotonin. In mammals, such chemical changes are associated with stress or anxiety. Following the shaking treatment, the bees were presented with a new, ambiguous odour – an equal mix of the two scents they had previously encountered. The goal was to determine how they would react. Would they optimistically extend their proboscis in anticipation of a reward, or would they avoid the risk of tasting bitterness? The unshaken bees responded randomly, extending their proboscis about half the time. In contrast, the shaken bees were significantly less likely to extend their proboscis, suggesting they had developed a temporary state of fear and were behaving pessimistically. Interestingly, this effect lasted only about five minutes before dissipating.

In a related experiment, Prof. Lars Chittka demonstrated with bumble bees that even a small amount of sugar water at the start of an experiment could induce a more optimistic approach to subsequent decision-making tasks. This suggests that bees, like other animals, can experience emotional states that influence their behaviour. A little treat goes a long way.

The brain and emotions

Experiments like these challenge the notion that bees lack emotions. While their brains are much smaller and structurally different from mammalian brains, some of the physiological processes within them are surprisingly similar to ours, or at least comparable. So perhaps they are capable of feeling emotions after all. However, in the eighteenth century, naturalists made the mistake of jumping to conclusions, interpreting the bee's waggle dance as an expression of joy upon discovering flowers, and we may be making similar errors today. About 150 years ago, Charles Darwin argued in *The Expression of the Emotions in Man and Animals* that all animals experience emotions. He pointed to behaviours like tail wagging, piloerection (the involuntary raising of hair or fur), and heavy breathing as evidence and suggested that even insects express emotions such as anger and fear. He also deduced that they experience envy and love – as their chirping could not be explained otherwise.

For animals to exhibit emotional states, they must display both specific behaviours – such as snarling in dogs or stinging in bees – and measurable physiological changes, such as elevated levels of certain neurotransmitters. Additionally, emotions require a degree of conscious perception by the animal, which is the most difficult criterion to prove. Some biologists even argue that it is impossible. The experiments by Bateson and Chittka explore this complex question, but many mysteries remain. Not all biologists are convinced that bees, or animals in general for that matter, experience emotions. Some consider the question irrelevant, arguing that it is impossible to determine definitively how any sentient being experiences the world, making it a matter beyond the reach of scientific inquiry. Others, however, propose that we should begin with the assumption that all beings share a similar subjective experience and focus on disproving this hypothesis instead. Prof. Tugrul Giray falls into that category when he says: 'We know that bees exhibit many characteristics of an aggressive state of mind when their nest is attacked. Why do we assume that this is the only emotion-like state bees have?'

Many mysteries still remain in exploring the emotions or otherwise of bees. From an evolutionary standpoint, emotions may act as algorithms to assess situations quickly.

Bees are originally forest dwellers. Tree hollows are their natural homes.

Summer

Aphids and other sources of sustenance

Flowers are not the only food source for honey bees. They can find other sources in forests, such as the honeydew produced by various species of aphids. Aphids pierce plant tissues – specifically the phloem – to access the nutrient-rich sap inside. Because the liquid in the phloem vessels is under pressure, aphids absorb more sap than they can process and excrete the excess as honeydew. This honeydew, which often accumulates on leaves, is a particularly rich source of food for bees, containing even more sugar than flower nectar.

TOP: Aphids primarily feed on the protein in plant sap, which they obtain by piercing the plant tissues.

ABOVE: They excrete the excess liquid, which is rich in sugar.

Making honey

Bees collect the honeydew and store it in their crop, a specialized structure in their oesophagus designed for transporting nectar or honeydew. The crop can hold up to 0.04 millilitres of liquid – approximately 40 milligrammes. While this might seem like a tiny amount, it is nearly 2/3 of the forager bee's body weight of approximately 60 milligrammes. Remarkably, bees can also carry large pollen packets weighing up to 20 milligrammes while flying. This extraordinary ability to transport their own body weight is a feat no modern cargo aircraft could replicate!

The crop is also a chemical marvel of nature where the magical transformation of honeydew or nectar into honey begins. Inside this specialized organ, enzymes break down fructose and glucose, while other compounds act as natural preservatives, ensuring the liquid remains stable and preventing spoilage. This initial processing sets the stage for the next steps, which take place back in the beehive.

Honey bees collect the sugary excretions of aphids, known as honeydew, and turn them into forest honey.

In the hive

In the hive, the forager bees deliver their liquid cargo to worker bees, which store it in honeycombs. However, the freshly collected honey initially contains too much moisture, with around 70 per cent water. In this form, the honey provides an ideal environment for bacteria, making it unsuitable for long-term storage. Therefore, bees fan their wings over the honeycombs for hours and heat the honey-filled cells, located at the edge of the capped honey area, to accelerate the evaporation of water. Once the water content drops to approximately 18 per cent, the honey cell is ready to be capped with a wax cover, creating a hermetically sealed cell. The combination of high sugar content and low water content is key to honey's remarkable shelf life. Additionally, fermentation gives the honey a slightly acidic pH, and enzymes contribute to its antimicrobial properties. For example, the enzyme glucose oxidase, introduced from the bee's honey stomach, continuously produces small amounts of hydrogen peroxide from sugars, giving honey its natural antiseptic qualities. This makes it an inhospitable environment for bacteria and fungi, ensuring its durability and purity.

A bee's crop can hold approximately 40 milligrammes of liquid, which she empties directly into a honey cell.

5,000-year-old honey

Honey stored in combs can remain perfectly preserved for decades. In fact, the oldest honey ever discovered was unearthed in Georgia in 2004 while excavating oil pipelines; it was an astonishing 5,000 years old. However, if honey is not harvested by humans, the bees will typically have consumed it by the following spring. During

TOP: By fanning their wings and heating the honey cells, bees reduce the honey's water content to 18 per cent; only then is it properly preserved, as no bacteria can survive in it.

ABOVE: Special enzymes in the bee's digestive tract further preserve the honey.

the summer, bees can bring up to 9 kilogrammes (20 pounds) of nectar into the hive daily. While stockpiling for the winter is their primary goal, there is another reason for this prolific storage behaviour that extends beyond mere survival.

Building a queen cell

The term 'queen bee' is somewhat misleading. In many languages, she is simply referred to as the 'mother bee', a name that better reflects her role. A beehive operates more like a democracy than a monarchy. In fact, US bee researcher Thomas D. Seeley describes a colony's social structure as a 'bee democracy', where many decisions are made collectively by the colony rather than being dictated by the queen. All power in the hive is vested in the people, in this case the worker bees.

When conditions are favourable, with the honey chambers full and the colony growing to an impressive size of 50,000 or more, the worker bees make a momentous decision: to build queen cells. This is often triggered by overcrowding in the hive. When thousands of bees live in close quarters, their constant contact signals the

When space in the hive becomes limited, the worker bees make a crucial decision. They begin producing special offspring by creating several queen cells.

need to split the colony. The first step in this process is the construction of special queen cells. These cells are larger than regular cells so that they can accommodate the larger body of a developing queen. In addition, they protrude at a right angle from the comb, making them easy to recognize even by a layperson.

Once the queen lays a fertilized egg in one of these cells, a cascade of complex events begins, ultimately leading to the division of the colony. This division, or swarming, is the only way honey bees reproduce – by founding new colonies. Bees have separated sexuality from reproduction. The sperm used by the queen to fertilize her eggs comes from a single nuptial flight during her first spring, when she mates with multiple drones. This event happens only once in her lifetime, yet it supplies her with enough sperm to produce offspring for the rest of her life.

Interestingly, the female offspring of the queen – the worker bees – are not reproductive and will not produce offspring of their own. For a new generation of bees to emerge, a new reproductive female – a queen – must be created. The workers ensure this by building queen cells and caring for the fertilized eggs inside. These steps ensure the colony's survival and the continuation of the species.

A sealed queen cell. A new queen bee is developing inside.

Since queens are longer than regular worker bees, their cells must also be larger and specifically longer to accommodate them.

Power food: royal jelly

As soon as the queen larva hatches – initially indistinguishable from the other larvae – she is fed royal jelly. Unlike regular larvae, which receive royal jelly for only a few days, a queen larva is fed it until she pupates. The royal jelly fed to future queens also has a slightly different composition, containing about 25 per cent more sugar. This enhanced sweetness plays a crucial role in determining which genes are activated in the larva's body, ultimately shaping its development. As a result, the larva metamorphoses into a larger, reproductive queen instead of an ordinary worker bee.

Swarming: leaving the hive behind

Meanwhile, the old queen is put on a harsh regimen by her worker bees, who relentlessly chase her through the hive, pinching and nipping her. This treatment serves a purpose: the queen must lose weight and build up her fitness, preparing for the major challenges that lie ahead. Then, something remarkable occurs in the hive, which still puzzles researchers to this day. About half of the bees decide to leave and form a new colony. The decision-making process – determining which bees will stay and which will go – remains a mystery. The result, however, is two relatively balanced swarms, with the youngest and oldest bees staying behind in the original hive. The bees preparing to leave exhibit distinctive behaviour; they bury their heads deep into the honey cells and fill their honey stomachs with as much honey as

TOP: During swarming, the old queen leaves the hive together with about half the bee colony.

ABOVE: As if responding to an inaudible signal, thousands of bees gather at the entrance.

possible. This is a vital step, as the departing bees will rely entirely on these provisions to survive for several days during their search for a new home. This meticulous preparation showcases the incredible coordination and resilience of the bee colony.

In search of a new home

As if in response to an imperceptible signal, the bees that are leaving stream towards the hive entrance in the thousands, gathering into a chaotic but purposeful mass. Then she appears: the queen. For the first time since her maiden flight, she leaves the darkness of the hive. If the colony lives in a harsh environment, with little food, this may have been four or five years ago; if the hive lives in a fertile environment, only a year may have passed. However long it has been, she is embarking on a completely new chapter of her life together with her colony. Accompanied by approximately 15,000 bees, she sets out in search of a new home.

The hive they are leaving behind is in prime condition for the next generation. It boasts thousands of cells filled with honey, pollen, and larvae, a vibrant workforce of young, healthy worker bees, and an experienced cohort of older bees who will guide the young and teach them the secrets of bee life. Several young, strong queens remain sealed in their queen cells, ready to emerge and compete for the throne. Whichever queen prevails will inherit a colony well equipped for survival and prosperity.

Many bees gather and swarm in front of the nest entrance before their departure.

TOP: During swarming, thousands of bees embark on a journey to establish a new colony.

ABOVE: The bees' only reserve is the honey that they have consumed and stored in their crops before leaving their old nest.

The first scout bees quickly set out to search for a suitable new home.

After departing, the old queen and her entourage settle nearby, typically on a branch or under an overhang no more than 30 metres (100 feet) from the original hive. Here, they form a dense ball, similar to the one they create during winter to conserve heat, with the queen safely positioned at its centre. This stage of swarming is critical and precarious. Without a new home, the bees can survive only a few days on the honey stored in their crops. If the weather turns or the search for a new home takes too long, they are doomed. Only one in four swarms successfully survives swarming. Remarkably, swarming bees are exceptionally docile. With no hive to defend, the bees are focused soley on protecting their queen and seeking a new home.

Now begins one of the most exciting and critical phases in the life of the new bee colony: up to 300 experienced forager bees – called scouts – fly off in all directions in search of a new home. Which bees set off to find a home for the colony seems to be determined genetically. Studies suggest that this trait is inherited from the drone father as scouts are significantly more likely to have the same drone as a father.

The birth of a queen

Meanwhile, back in the bees' old home, a great spectacle is announced by an unfamiliar sound. A 'quacking' is heard in the beehive. Bees don't have ears, but they can still detect certain acoustic signals using specialized sensory organs. Their antennae are equipped with organs that allow them to perceive particle velocity, a physical quantity related to sound pressure. However, this allows them to hear sounds only in their immediate vicinity. But, like all insects, they possess another organ with which they can perceive sound waves. The subgenual organ located in each of their legs allows bees to detect sound waves that cause the ground they

A completely harmless swarm of bees rests together, focused solely on staying close to their queen.

Several dozen experienced forager bees stop searching for colourful flowers and instead begin looking for dark tree hollows to find a new home for their colony.

are standing on to vibrate. Fortunately, when they are inside the beehive, bees are usually standing on a comb, and this structure is particularly good at conducting vibrations. This is why Prof. Jürgen Tautz also calls it the 'bee telephone', as it facilitates a wide range of communication throughout the entire colony.

The queen produces the strange 'quacking' sound by vibrating her flight muscles in her queen cell. The sound is probably intended to signal to the worker bees nearby

that the queen is hatching. Next, the queen practically saws off the lid of her queen cell with her proboscis. She repeatedly stretches her 'tongue' through one of the holes so the worker bees can feed her honey to keep her strength up.

The new queen

After hours of effort, the moment finally arrives: the new queen manages to open her cell like a tin can and emerges into the hive. Once she has hatched, the queen announces her presence to the entire colony with a distinctive sound known as 'piping', also referred to by beekeepers as 'tooting'. The queen produces this sound,

TOP: Shortly after the swarm has left the old nest, the new queen emerges.

ABOVE: The queen uses her mandibles to 'saw' her way out of the queen cell from the inside, leaving a characteristic cut at the tip of the cell.

which is unique to this extraordinary moment, by leaning against the comb, uncoupling her flight muscles from her wings and vibrating them.

If the hive still has ample supplies and a large enough workforce, the worker bees may send a second swarm into the world, led by this newly hatched queen. However, a perilous fate awaits the remaining unhatched queens still in their cells. The worker bees actively drive the new queen away from these cells to prevent her from committing a seemingly ruthless act: Newly hatched queens instinctively seek out their unhatched rivals and attempt to kill them with their stingers while they are still in their cells.

If the colony is not large enough or there are not enough resources, the worker bees will not protect the other queen cells, allowing the first queen to hatch and eliminate her rivals. The new queen moves throughout the hive, systematically killing her sister queens before they can hatch. Occasionally, two queens may hatch at the same time, leading to an immediate fight to the death. Only one queen will survive to lead the colony.

The seemingly brutal practice of creating multiple queen cells serves an essential purpose. Not all queen larvae develop successfully, and the colony cannot afford to risk being without a queen for too long. By producing multiple queen cells, the bees ensure the survival of the colony, with the strongest queen ultimately prevailing – unless a beekeeper intervenes.

The queen opens the cell and steps out of her nursery.

Occasionally, multiple queens hatch at the same time, but they immediately face off in a fight to the death.

The nuptial flight

Just a few days after hatching, the queen departs on a nuptial flight, but she relies on her sisters to guide her, as only they know where the drones from other colonies gather. Following their lead, she mates with up to twenty drones, ensuring a high level of genetic diversity in her offspring. This diversity is crucial to the colony's success, as it influences traits such as pollen and nectar preferences, scouting behaviour, and other essential instincts, which are largely inherited from the father.

Upon returning from her nuptial flight, the queen begins laying eggs, which can either be fertilized, from which female bees will emerge, or unfertilized, from which drones will hatch. The worker bees decide which type of eggs to lay, not the queen. If the colony requires drones, the workers build slightly larger cells. As the queen inserts her abdomen into a cell, she instinctively senses its size and lays an unfertilized egg in the larger cells and a fertilized egg in the smaller cells.

The queen's ability to lay fertilized eggs depends on the sperm stored from her nuptial flight, which can last up to five years. Once this supply runs out, she can only produce drones. The worker bees quickly detect this change, which marks the end of the queen's reign. Despite her potential to live longer, the workers create new queen cells and eventually replace her. The old queen is cast out, and her time in the hive comes to an end.

Far from being a ruler, in reality the queen is the most controlled and subordinate member of the colony. Her title, though grand, does little to reflect her true role as a tireless egg-layer entirely dependent on the worker bees for her survival and continued relevance within the hive.

The old queen continues her search for a home

In the wild, experienced scout bees search tirelessly for a new home for their displaced colony, switching effortlessly from hunting for bright flowers, which they have done all their lives, to finding the dark entrance to a tree hollow. We are so familiar with bees' collective behaviour that we often overlook how rare it is in nature for a single individual to take on the monumental task of finding a suitable home for tens of

A scout inspects a tree hollow that could serve as a home for her swarming bee colony.

thousands of relatives. Apart from ants and termites, no other creatures make such a communal effort; most species seek out or build shelter solely for themselves and their offspring.

Several dozen scout bees spread out over a radius of up to 5 kilometres (3 miles), searching for a suitable site for the swarm. Honey bees are originally forest dwellers and depend on tree hollows to live in the wild. They like to take over abandoned woodpecker nests, but such old, hollow trees have become increasingly scarce in modern landscapes.

When a scout bee discovers a potential entrance to a tree hollow, she approaches cautiously, checking to ensure it is uninhabited before daring to land, and then climbs inside to inspect the cavity thoroughly. From a bee's perspective, a good hollow must fulfil several criteria. The entrance should ideally be at the bottom, which makes it easier to regulate the temperature by preventing warm air from escaping. It should also be small enough to deter predators but not so small that it is difficult for the bees to enter. Issues like dampness or drafts are less concerning, as the bees can seal leaky areas quickly with propolis (see p.126).

The hollow must also be spacious enough to accommodate a growing colony. While the current swarm may consist of around 15,000 bees, their population could double or even triple in size. To assess the size, the scout measures the interior by pacing out the inside of the cavity in tiny steps. She also checks the height of the entrance from the ground, favouring locations at least 1.5 metres (5 feet) high, and ensures the entrance faces southeast or south so that the sun warms the colony early in the day. Based on this meticulous process, which highlights yet again the incredible adaptability and resourcefulness of honey bees, the scout bee ensures that the new home will support the colony's survival and growth.

Swarm intelligence

When a scout bee finds a suitable hollow, it flies back to the swarm. The bee then performs a waggle dance (see p. 61) in an attempt to persuade her sisters to visit the site, much like a forager bee that has located a rich source of flowers. The dance conveys approximate information about the direction and distance of the potential nesting site. However, she is not the only scout with news to report. Other scouts also return from their exploratory flights, each with its own discoveries. Some might be covered in soot particles from searching chimneys and others in dust after investigating attics. Unlike forager bees, none of these scouts have pollen on their legs, as their sole mission is to find a new home, not food.

A scout tries to determine if a tree hollow is uninhabited before entering it.

But how do these scouts communicate their findings, and who decides where the swarm should go? What follows is one of the most remarkable displays of collective decision-making in the animal kingdom, a process so impressive that it inspired the term 'swarm intelligence'. Each scout bee that believes she has found a suitable site dances to promote it, competing indirectly with other scouts advocating for different sites. At first, each bee shares her discovery with only the sisters immediately around her. These recruits then fly off to inspect the proposed site for themselves, forming their own opinion. Meanwhile, the original scout bee repeatedly revisits her site, marking it with pheromones from the Nasonov's gland on her abdomen to guide others.

If the new inspectors are impressed, they return to the swarm and begin dancing in support of the same site, gradually boosting the site's popularity. Through this iterative process of evaluation and recruitment, the swarm collectively converges on the best location, guided by the combined wisdom of the scout bees.

The entrance is evaluated to ensure it faces the optimal direction, ideally south.

A scout bee moves through the interior of a tree hollow, measuring its size. It must be spacious enough to accommodate a colony that could double or even triple in size.

Deciding on the new nesting site

The bees' decision-making process for choosing a new site has fascinated generations of biologists. Thomas D. Seeley, Martin Lindauer, Jürgen Tautz, and many others have helped to gradually unravel the mystery of how bee swarms reach their collective decision. The process can last for days.

Bees may sometimes put their discussion on hold at night, only to resume from where they left off the next morning. Over time, more bees visit the potential nesting sites, forming their own opinions about each location. Depending on the duration of the decision-making process, after one or two days as many as a hundred bees may dance in support of a handful of possible sites.

As well as attempting to recruit additional bees to inspect the location themselves, scout bees may actively try to stop rival dancers. This can sometimes get quite physical, with bees grabbing the heads of a supporters of a rival site and giving them a brief shake to discourage further dancing. In the vast majority of cases, more and more bees gradually give up dancing for 'their' destination and join other factions in the bee colony. Remarkably, by the time a consensus is reached, fewer than five per cent of the swarm may have actually visited the chosen site. Yet, through this iterative process, a collective decision emerges.

If the site meets all the criteria, the scout bee decides to return to the colony to report her discovery.

Once all the scouts are dancing in favour of the same destination, the swarm enters its final phase. A call to action begins as scout bees make piping sounds, similar to those made by queens in the nest. This noise, produced using their flight muscles, spreads through the swarm. Approximately 30 minutes before departing, the piping grows more frequent, which signals the swarm to prepare for the journey to their new home.

Piping hot bees and buzz-runs

The piping sounds made by the scout bees trigger every bee in the swarm to raise her body temperature. This collective heating warms the entire swarm to 35–37°C (95–98.6°F), preparing them for takeoff. Buzz-runs signal the final 'go'. Scout bees, having already triggered the temperature increase, indicate that it is time to leave. What follows is an astonishing coordinated action. The swarm virtually explodes and hovers in place in the air, held together only by various scent signals emitted by the queen and the worker bees. Scout bees begin to fly back and forth between the swarm and the new home and, slowly but surely, the swarm extends in the direction it is supposed to fly. Gradually, more and more bees catch sight of their new home. After just a few minutes, the entire swarm reaches the new nesting site, where the hard work of building a new home begins.

Building a new home

For the bees, it is late in the year. Winter is only a few months away. Yet they have to start from scratch. Their nest contains no honey, no pollen, and no offspring. To survive the approaching winter, the bees must collect 1 kilogramme (1.2 pounds) of pollen and 2 kilogrammes (4½ pounds) of nectar each day; otherwise, their provisions will fall short. But first, they need to build combs.

Handmade wax

Like all insects, bees have wax glands. The wax secreted by these glands coats the exoskeleton, making it waterproof. This is not intended to protect them from the rain; rather, it prevents dehydration. In chemical terms, chitin, the primary component of an insect's exoskeleton, is a sugar, which means it is permeable to moisture. Without the layer of wax, insects would lose internal moisture to the environment and dry out. Bees, however, have taken the use of their endogenous wax to an entirely new level. Their wax glands are concentrated on the underside of their abdomen and produce far more wax than any other insect.

When swarming bees settle into a new home, the wax glands on their abdomens reawaken after weeks of dormancy. Normally, only young worker bees have active wax glands as their role is to build the hive's cells. In older forager bees, like those that travelled with the swarm, these glands are typically inactive. But in times of urgent need, such as building a new nest, bees demonstrate remarkable adaptability. Even older bees can reactivate their wax glands to contribute to the construction. Similarly, they can restart the glands that produce royal jelly to feed larvae, despite these glands also becoming inactive with age. Once again, bees reveal their extraordinary capacity to adapt and respond to their colony's needs.

Specialists and generalists

The combination of flexibility and specialisation in honey bees is unparalleled in the animal kingdom (with the possible exception of humans). Ants and other social insects feature highly specialized individuals within their colonies, divided into castes. For example, 'soldiers' in ant colonies are physically adapted to defend the nest with enormous mandibles, but their specialization limits them – they are incapable of performing other tasks within the colony. Conversely, other social animals, such as primates, wolves, or meerkats, practice task allocation, but their members remain generalists. Tasks are not assigned for extended periods, and individuals can switch roles as needed.

Bees, however, strike a unique balance between specialization and flexibility. Some bees spend days at a time focussed on specific tasks, such as caring for the brood, building combs, or collecting pollen. Nevertheless, they retain the ability to switch roles when necessary, even reactivating bodily functions, such as the glands that produce royal jelly or wax – essential components of the beehive.

Bees literally create their own world. While humans extract building materials, like wood and minerals, from nature, bees produce their own in the form of wax. Beeswax is an extraordinary substance, composed of over 300 individual chemical components, and bees can detect its exact chemical composition. Each bee's wax contains a unique chemical fingerprint, formed by a few substances present in minute quantities. These trace elements are determined genetically and enable bees to recognize how closely they are related to one another. For example, a bee can

All insects have wax glands, but bees are the only insects that have evolved to use their wax as a building material.

Freshly built wax combs are snow-white, gradually darkening over the years.

differentiate between a half-sister (sharing the same mother but a different father) and a full sister (sharing both parents) based on the chemical composition of their wax. Due to their genetics, full honey bee sisters share more genetic material than full human siblings. By interpreting these subtle chemical cues in the wax, bees can navigate the complex social dynamics of their colony with remarkable precision.

A perfect building material

As if that were not enough, bees also have the remarkable ability to alter the composition of their wax to adjust its static properties as needed. They can modify it to harden more quickly or remain supple for longer, depending on their requirements. However, wax production requires a great deal of energy: bees need to consume 6–7 kilogrammes (13–15½ pounds) of honey to produce just 1 kilogramme (2.2 pounds) of wax.

The construction process begins with the bees creating a scaffold using their own bodies. By hooking onto each other's legs, they form a living net that allows other bees to climb up and down. At the ceiling of the hive, they attach the first points of the comb, which then grows downwards as a free-hanging structure. Unlike managed hives where beekeepers provide frames, wild bee colonies create stunning organic, curved formations. These structures, aligned with Earth's magnetic field, are among the most beautiful in nature. At the constant temperature the bees maintain in the nest, beeswax is both supple and sturdy, perfectly suited for housing the brood and storing honey and pollen. The blend of functionality, adaptability, and artistry in their construction is a testament to their incredible engineering skills.

Architectural masterpieces

Freshly built honeycombs range in colour from pale yellow to nearly white. They darken over time, first turning the characteristic orange of beeswax, before eventually turning a deep brown. Honeycombs are architectural masterpieces, but their iconic hexagonal shape is not the result of the bees' mathematical skill as previously believed. Instead, physics is responsible. The design allows bees to maximize the number of cells while using the least possible amount of wax. It comes as no surprise that human engineers often mimic the structure of honeycombs to create lightweight yet robust designs.

Natural combs, free from man-made frames, have an organic design.

If the entrance to the nest is too large, the bees make it smaller using wax and propolis.

Initially, bees construct the cells in a circular shape to match the contours of their own bodies. As Prof. Jürgen Tautz and his team have demonstrated, a heater bee then enters the finished cell, warming it enough to soften the wax. Once softened, the wax responds to the surface tension, naturally forming the hexagonal shape. A similar phenomenon can be observed with soap bubbles that cluster together, creating flat, straight partitions between them. If soap bubbles were aligned on a single plane, the hexagonal structure of a honeycomb would emerge. Once the wax cools and solidifies, the cells retain their perfect hexagonal shape.

The blind beekeeper

Swiss naturalist François Huber made a remarkable discovery about bees over 200 years ago. Blind himself, he conducted his experiments with the assistance of his wife, Marie-Aimée Lullin, and his servant, François Burnens. Despite his disability he was able to make many astonishing discoveries. His first significant discovery was that the queen bee does not mate inside the hive, but instead flies out and mates in the air, often high above the ground. He also discovered that queen bees develop from the same eggs as normal bees. To study how bees construct their combs, Huber and his assistants replaced the hive cover with a glass pane, allowing Marie-Aimée and François to observe the colony's behaviour. Typically, bees begin by attaching the first cells to the top of the hive, using gravity to guide their downward construction. However, when confronted with a glass pane, the bees adapted by flipping their approach entirely – attaching their combs to the bottom of the hive and building upwards instead.

The experiment did not stop there. Huber and his team also replaced the bottom of the hive with a glass pane. This time, the bees built their combs horizontally, spanning from one wall to the opposite side, creating a bridge-like structure. The most striking part of the experiment came next. After the bees had constructed their comb halfway, the naturalists removed the opposite wooden wall and replaced it with yet another glass pane. Now, the bees had no endpoint for their cantilevered structure. In response, they adjusted their strategy, bending their comb at a 90-degree angle towards the closest wooden wall.

The bees' reluctance to attach combs to glass is well-founded. Wax does not adhere well to such surfaces, which rarely exist in nature. However, their ability to completely reverse their construction method demonstrates extraordinary cognitive flexibility. Their capacity to adapt their plans when the anticipated goal changes is even more impressive. The bees not only recognized that the endpoint of their construction was no longer viable but also communicated this information throughout the colony. Together, they redirected their collective efforts to build towards a new destination. How they coordinate such a shift in strategy remains one of the many mysteries of bee behaviour.

The first wax cells are anchored at the top; from there, the comb is built straight downwards.

Propolis: a remarkable sealant

Bees seal holes or cracks in the tree hollow using propolis – another building material with remarkable properties. Some of the worker bees specialize in collecting the soft, resinous coating of young tree buds in the afternoons, transporting it back to the nest on their hind legs, much like pollen. They typically gather resin from birch, poplar, chestnut, or willow trees. In nature, the resin protects young shoots by inhibiting the growth of bacteria, viruses, and fungi. Bees mix the resin with their saliva to create propolis, which has antibiotic properties primarily derived from the plants. In the hive, propolis serves a similar function to that of resin on tree buds: it offers protection against fungi and bacteria. Bees coat the edges of comb cells with propolis both to stabilize them and, more importantly, to shield the brood from infection. Over time, they cover the entire interior of the tree cavity with a protective layer of propolis, thus enhancing the hive's defences against disease.

Dealing with parasites

In Europe and North America, it has almost been forgotten that bees can survive without the help of beekeepers. Since the introduction of the *Varroa* mite from Asia to Europe, it has been widely considered impossible for bees to survive without protection from this parasite. The *Varroa* mite invades the brood cells of bees, feeding on the larvae's hemolymph (the insect equivalent of blood, which, while different in composition, serves similar functions). Once the bee emerges from its cell, the mite remains on its host, continuing to feed on its bodily fluids. In the process, the mite transmits various infectious diseases, ultimately weakening and killing the bee. Since the arrival of the *Varroa* mite in Europe, most beekeepers spray their colonies twice a year with chemicals such as formic or lactic acid to control infestations. However, subjecting bees' sensitive sensory organs to acidic substances is an unpleasant practice for any beekeeper. Despite this, the majority of beekeepers, whether professionals or hobbyists, believe that bees cannot fend off the *Varroa* mite on their own.

Even in natural nests, guard bees are responsible for ensuring that no foreign bees are allowed to enter.

The decline and recovery of bee populations

Some beekeepers, however, are exploring alternative approaches. Roland Sachs, a civil engineer who dedicates much of his spare time to working with bees, aims to raise bees without treating them against *Varroa* mites. His strategy centres on minimal intervention, allowing nature to take its course. Sachs draws inspiration from beekeepers in the UK who have successfully implemented similar methods. His efforts suggest that stepping back, rather than stepping in, might be the key to fostering resilient bee colonies.

In 2006, a small group of beekeepers in Wales decided to stop treating their colonies against *Varroa* mites. During the first winter, they suffered dramatic losses, but some colonies survived and thrived the following summer. Instead of relying on 'purebred' queens from reputable breeders, they chose to work exclusively with locally adapted bees. This approach included recapturing swarming colonies rather than importing new ones. Surprisingly, over the years, their bee populations gradually recovered. Colony losses stabilized, honey yields remained consistent, and the bees became more resilient.

It appears that, without direct intervention, they had effectively 'bred' a hardy, locally adapted strain of bees. Instead of artificially selecting traits, they allowed nature to take its course, leaving the process to natural selection. Bee researcher Peter Neumann coined the term 'Darwinian beekeeping' to describe this approach, inspired by Charles Darwin's principles of natural selection. The strategy has proven successful: today, large areas in Wales are home to bees that no longer require treatment against *Varroa* mites. These populations have developed a natural tolerance against the parasite.

Similar efforts are taking place in Germany, the Netherlands, Poland, Norway, the United States, and other countries, where some beekeepers are trying to abandon chemical treatments against *Varroa*. However, this approach is practised by the minority and faces resistance from followers of traditional beekeeping methods and beliefs. So while there is a growing trend towards more natural beekeeping, it has yet to gain widespread acceptance.

Wild bees

A potential model for more natural beekeeping practices could be the wild honey bee populations that are being rediscovered in various regions of Europe. Although humans have kept and used honey bees for thousands of years, they have not fundamentally altered them through domestication, as has been the case with other livestock. There are no significant genetic differences between wild honey bee populations and those managed by beekeepers. However, in Europe, there is still a lack of extensive, systematic data on the distribution and prevalence of wild honey bee populations. Fortunately, dedicated enthusiasts have begun locating and documenting these wild colonies in databases, contributing valuable insights.

Wild colonies, much like their ancient ancestors, often inhabit natural forests, where they use empty tree hollows as nests. Their lifestyles differ significantly from those of colonies managed by beekeepers. For instance, wild colonies are free to swarm regularly, leading to periods when they have no brood, which makes them less susceptible to the *Varroa* mite and other parasites. Additionally, the limited space in tree hollows often keeps these colonies smaller than those in managed hives. Natural nesting sites in forests are also spaced far apart, which greatly reduces the risk of diseases being transmitted between colonies.

Natural combs darken over the years, gradually taking on a rich, deep colour with prolonged use.

These factors make wild colonies less vulnerable to parasites, offering a plausible explanation for their continued existence in Europe, something once thought impossible. The resilience of these wild honey bees highlights the potential benefits of allowing colonies to live in ways that align more closely with their natural behaviours.

Symbiotic relationships

Living in a natural tree hollow is believed to offer numerous benefits for bees, many of which are only now being studied. For example, wild bees appear to live in symbiosis with other organisms inhabiting the same tree hollow. One such creature is the pseudoscorpion, or book scorpion, a tiny arachnid just a few millimetres in size, which seems to help bees by preying on the *Varroa* mites that infest bee larvae. Another example is the wax moth, which, though unwelcome in managed hives, lays its eggs in abandoned combs of wild nests. Its larvae then effectively clean up the old nest, removing parasites and making the hollow ready for a new generation of bees to settle. Despite thousands of years of beekeeping, much remains to be discovered about the natural behaviour of bees and the intricate ecological relationships they form in their wild habitats.

Natural air conditioning

After settling into the tree hollow and constructing its first combs, the colony faces its next challenge: unusual weather patterns brought about by climate change, such as altered seasons, heavy rainfall, and extreme heat waves. Although honey bees originally evolved in the tropics, heat waves are a major stressor for a colony. To protect its larvae, the temperature inside the hive must not exceed 36°C (96.8°F). Prolonged exposure to temperatures that are too high or too low can cause severe developmental issues or even lead to the death of larvae and pupae. Ensuring this delicate balance is critical for the colony's survival and future growth.

The principle of ventilation

To lower the temperature inside a tree hollow, bees employ a strategy they have been using for millions of years – a method humans have only recently begun to understand. The process begins with a group of bees assuming the role of 'fanner bees'. These bees position themselves both at the entrance and within the nest, vigorously flapping their wings to generate a flow of air. They arrange themselves in a precise formation, maintaining a uniform distance from one another to create the strongest possible ventilation. This flow of air pushes warm air from inside the nest to the outside, helping to cool the hive.

TOP: Flight muscles are used for more than just flying; when temperatures rise, bees use them to fan and cool the nest.

ABOVE: When fanning, the bees position themselves in a formation designed to maximize airflow efficiency.

If fanning alone is not enough to reduce the temperature, the bees take additional measures. Scout bees search for water, relying on their keen eyesight to spot the glimmer of a water surface, as they cannot detect water through smell or sound. Once a suitable source is found, the scout bees alert their sisters, who begin collecting droplets of water and transporting them back to the nest in their crops. On their return, the bees carefully deposit the water droplets near the brood cells. The fanner bees continue their work, circulating air through the nest. As warm air passes over the water droplets, the resulting evaporation creates a cooling effect, essentially mimicking the principle of air conditioning. This evaporative cooling can lower the nest's inside temperature by several degrees Celsius, ensuring that the maximum brood temperature of 36°C (96.8°F) is not exceeded, even during midsummer heat waves.

Bees need water not just to quench their thirst but also to help cool their nest.

Clearing out bad air

When attempting to improve the quality of the air inside the nest, the individuality of the bees comes to the rescue again. François Huber, ever curious about bees' mysterious ways, became particularly intrigued by how they regulate the hive's temperature and improve air quality. With the help of his assistants, he discovered that, as the air quality worsened and oxygen levels dropped, more and more bees would stop whatever they were doing to flap their wings. When the hive became especially stuffy, all the worker bees would collectively set aside their tasks and focus on ventilating by fanning their wings.

Huber proposed an elegant theory. He suspected that each bee reacted differently to harmful smells. The most sensitive bees would start fanning first and, as the situation deteriorated, the less sensitive ones would join in. In this way, without any centralized control, the colony as a whole managed to ensure proper air circulation. It is a beautifully simple system, but Huber could not test his hypothesis at the time as his team lacked the tools to mark individual bees.

Still, he was onto something. Modern experiments have since confirmed that honey bee colonies adopt a flexible approach to organizing essential tasks inside the hive. Each bee seems to have its own 'opinion' about ideal air quality, temperature, or even the size of nectar and pollen reserves. They react accordingly and, when a problem becomes urgent, more bees step in to solve it. It is not perfect, but that is the magic: the collective thrives because each bee responds in its own way.

The bee and its fearsome sting

Although our fear of being stung by a bee is quite understandable, it is nevertheless important to know what a stinger is all about. It evolved millions of years ago from a body part on the abdomen that was originally used by insects to lay eggs – the ovipositor or 'egg-laying device'. Some insects, such as the grasshopper, still use this ovipositor to lay eggs. The fact that the stinger evolved from this part of the female body also explains why male bees, the drones, do not have a stinger. The stinger has evolved so that bees can defend themselves against their worst enemy: other bees. Towards the end of summer, when the beehive is filled with honey, it can be easier for bees to pillage another bee colony than to gather nectar themselves and process it into honey. For this reason, bee colonies often raid each other's hives during this time of year. First, individual scout bees fly to the nest to assess the size of the colony. If the colony appears small, and thus weak, they attack.

The stinger evolved from the ovipositor, a tubular body part used by many insects to lay eggs; that is why only female bees have a stinger.

War between bee colonies

One species of honey bee, the Africanized bee (*Apis mellifera scutellata*), can be especially dramatic. Africanized bees originated from breeding experiments in Brazil during the 1950s. They attempted to cross-breed European bees with honey bees imported from Africa to achieve higher honey yields. Whilst the attempts were successful, the new breed of bees they created was particularly aggressive and spread across the entire continent to North America. This subspecies even takes over foreign bee colonies by infiltrating their nests, killing the queen, and replacing her with their own.

Sometimes, these battles last for hours or even days. Because these conflicts between thousands of individuals of the same species are fought over resources – honey that is – and not over competition between mating males, and because the battle is fought to the death, these conflicts could be described as wars between colonies. Furthermore, the outcome of these conflicts is less determined by the strength of individual bees than by the ability of the colonies to supply the bees on the 'frontline' or at the entrance to the nest with food. Supply routes are therefore more crucial than the strength of the individuals. Nevertheless, war metaphors are perhaps misplaced when it comes to bees. The conflict has nothing to do with the ethical and psychological dimensions of human warfare. In the end, it is more about

Bees do not 'sacrifice' themselves when they sting because they do not know that it will kill them. In the course of evolution, the stinger developed as a weapon against other insects, which they could use without dying.

efficiency, as it is more efficient for these insects to rob each other of food than to gather it for themselves.

The ability of bees to collaborate – their impressive collective performance – has always served as inspiration for ideologies. Some see the honey bee as an ideal from nature for collective organizations. Even the ancient Egyptians recognized the exceptional nature of the bee and made it a hieroglyph representing the Pharaoh (together with the hieroglyph of the papyrus stem, which was used for writing). Napoleon adorned himself with stylized bees on his royal cloak. Alongside this fascination with the extraordinary collective achievements of bees, however, there soon emerged the myth of the extraordinary self-sacrifice bees make for their colony – after all, a bee dies when she stings. However, this interpretation is misled. A bee does not sacrifice herself for her colony when she stings. There is simply no mechanism to prevent her from dying when she stings a mammal. The insect stinger evolved over 140 million years ago as a defence mechanism for insects against other insects. In this phase of Earth's history – the Early Cretaceous Epoch – there were only insects and reptiles on land. Mammals evolved later. Bees used their stinger to defend themselves against other insects, and at the time, they could use it without the stinger getting stuck in the victim and dying when it was removed. Later, as some bee species formed colonies, they no longer defended just themselves but also their offspring or the food reserves in the nest. Only then did the first mammals appear, with their characteristic layer of fat under the skin. The stinger gets

Honey bee venom has more than 50 different components. Aside from the main toxin, it consists of enzymes that break down cell membranes and connective tissues to amplify the effect of the toxin.

stuck in this layer of skin, and when the bee attempts to withdraw it, the stinger is torn off along with a large portion of her abdomen, which proves fatal for the bee. As bad as this death is for the individual bee, it plays no role in genetic selection among bees. Therefore, there has been no change in the anatomy of the bee stinger over the course of evolution, and the notion that bees defend their colony as an act of self-sacrifice is simply inaccurate.

Taming killer bees

The term 'Africanized honey bee' often conjures up scenes of chaos. These insects, notorious for their hyper-aggressive behaviour, have claimed many lives across Central and South America. In Brazil, a beekeeper stumbled too close to a hive and was stung over 1,000 times. In Costa Rica, a farmer attempting to clear vegetation away from a nest inadvertently unleashed an attack that left several of his livestock dead. Death tolls in some Central American countries have climbed into the hundreds as these 'killer bees' have relentlessly defended their hives, stinging anything within range.

Just a handful of queen bees escaped from the experimental apiaries in Brazil in 1956, but they began to breed with local honey bee populations, giving rise to a hybrid bee with remarkable resilience and a propensity for aggressive behaviour. By the 1980s, these bees had spread throughout Brazil and the rest of South America, crossing into Central America before reaching the southern United States by the early 1990s.

But something unexpected happened when Africanized bees arrived in Puerto Rico in the 1990s. Unlike their mainland counterparts, which remained notorious for their aggression, Puerto Rico's bees became much more tame within a period of 30 years. Africanized bees went from living up to their menacing nickname to becoming one of the most docile bees in the world. Beekeepers report working without protective suits and accessing hives with minimal protest, as European beekeepers are accustomed to. Nevertheless, DNA analysis confirmed their identity: these were still Africanized honey bees but with a fundamentally different temperament. This raised an intriguing question: what had happened on the island?

Enter Prof. Tugrul Giray, a researcher at the University of Puerto Rico, who sought to unravel this mystery with his team. Their findings revealed a fascinating interplay between genetics, natural selection, and environmental pressures. Africanized bees in Puerto Rico were faced with the special conditions of a tropical island environment. A wet and a dry season result in a flower pattern similar to temperate regions. Flowers are available for only part of the year, while there are almost no flowers for roughly half of the year and thus almost no food for the bees. Additionally, Puerto Rico is an island. If the flowers are not blooming, wild bees cannot move on and look for a better location. They are trapped on an island.

Although bees are not as specialized as ants, bees that protect the nest do not forage and forager bees take much longer to respond to an intruder and do not respond with the same level of aggression as dedicated 'soldier' bees. As a result, bee colonies had to 'decide' if protecting their nest or dedicating as many bees as possible to finding food was a better strategy. The Puerto Rican bee colonies experimented with both strategies, and evolution favoured the strategy that was more successful. Prof. Giray documented that more docile colonies foraged more effectively, stored more honey, and reproduced more successfully. In the course of only three decades, natural selection favoured the more docile bees, leading to a population-wide shift in behaviour.

Giray's team uncovered genetic evidence to support this theory. The bees' genomes showed they were still overwhelmingly African in origin, but small changes were observed in genes associated with behaviour, suggesting rapid evolution. The key lies in Puerto Rico's unique ecology. Its pronounced wet and dry seasons create a periodical nectar and pollen famine, a challenge similar to that faced by temperate honey bees, which must store large reserves of honey to survive through harsh winters. Puerto Rican Africanized bees had to adapt by prioritizing foraging over aggressive defence. Being trapped on an island amplified these pressures. With no opportunity to migrate or abscond in search of better conditions – a survival strategy common among Africanized bees on the mainland – the Puerto Rican bees faced an evolutionary bottleneck. Colonies that invested too much energy in defending their nests at the expense of foraging did not produce enough honey to survive the lean months. These colonies fared poorly, leaving behind more docile, more resource-efficient colonies that focussed on foraging. Over time, natural selection systematically favoured this docile temperament.

What happened in Puerto Rico exemplifies the profound impact of the environment on evolution. In less than three decades, Africanized honey bees – feared as 'killer bees' – evolved into ideal partners for beekeepers, a transformation driven by the need to survive in an island ecosystem. Giray's work not only highlights the resilience and adaptability of honey bees but also opens doors to understanding how environmental pressures can shape behaviour and genetics in other species.

Hornets: honey bee predators

In late summer, hunting season begins. Spectacular aerial battles often unfold in front of a beehive's entrance, where hornets swoop in to pluck individual bees out of the air. Ironically, the hornets cannot consume the bees they catch, as they lack the ability to chew the meat. Instead, they feed them to their larvae. The larvae, in turn, regurgitate part of the bees undigested, providing a share of the prey for the adult

To defend their hive from hornets, honey bees bite, sting, and – most impressively – swarm the intruder in a tight ball. Together, they raise the temperature with their body heat until the hornet overheats and dies.

hornets. Over 100 million years ago, the evolutionary paths of wasps (which include hornets) and bees diverged. While hornets remained carnivores, bees adapted to meeting their protein needs with pollen. Hornets have long had a bad reputation. Persistent myths, like the claim that three hornet stings can kill a person, only add to their undeserved infamy. In Europe, however, hornets hunt such a small number of bees that they do not pose a real threat to the survival of a honey bee colony. Indeed, the European hornet has become so rare that it is now listed as a specially protected species, and its presence near a hive is a positive sign – an indicator of a healthy, diverse ecosystem.

And it is not as if honey bees are defenceless against hornets. They have developed some truly fascinating strategies. At times, worker bees gather at the nest entrance to form an impressive blockade, thwarting hornets' attempts invade the hive and steal larvae. Even more remarkably, bees have been observed forming a tight ball around a hornet and heating up. In this dense cluster of bee bodies, the temperature rises to 46 °C (114.8 °F) – just one degree higher than a hornet can survive. The heat, combined with the restricted airflow, incapacitates the predator.

Like humans, bees loose their hair with age. Some develop tiny bald spots on their backs.

All for a drop of honey

Summer bees typically live for about six weeks, though it is hard to calculate this exactly. Most bees do not die in the hive but on their foraging flights, typically due to accidents or predators rather than old age. Yet, colonies often contain what are known as 'Methuselah bees', individuals that defy the norm and live for seven, eight, or even nine weeks. A bee's age can be determined based on her wings. Over time, the thin wing membranes fray, and eventually the wings fail. At up to 270 wing beats per second, the wings endure an incredible strain, moving through the air faster than the speed of sound. Bees essentially 'break the sound barrier' repeatedly – at least with their wings – which wears down the edges. As they age, their fur also lightens and small bald spots appear.

Calculating how much nectar a bee collects in her lifetime is not straightforward because the numbers vary so much. A bee can carry between 20 and 40 milligrammes of nectar in her honey stomach per trip. On some days, she might make just three flights; on others, she could make ten. And while some bees forage for 20 days, others only manage ten. This means a single bee might gather anywhere between half a gramme and 8 grammes of nectar in her lifetime. On average, this amounts to about a drop. And for that drop, the bee flies up to 800 kilometres (500 miles).

TOP: Flight muscle strength also decreases with age. An old bee won't be able to fly as far as a younger bee.

ABOVE: On average, summer bees live for approximately six weeks, while winter bees can live for almost six months.

Fog announces the arrival of autumn in the Karvendel mountains in Austria.

Autumn

A new generation, a new cycle

By midsummer, the first bee of the year has long since died, but the circle of life within the hive continues. As long as the weather remains warm, the bees tirelessly gather pollen and nectar. In Europe, ivy produces the final blossom of the year, providing essential sustenance for the colony. The queen continues to lay eggs, although in decreasing numbers as the year progresses, and the colony's population gradually shrinks to about 10,000 members. Towards the end of summer, as temperatures drop and food supplies dwindle, the bees make a crucial adjustment: they allow their offspring to hatch at a slightly lower brood temperature – 34.5°C (94.1°F) instead of the usual 36°C (96.8°F). By October, it is typically too cold for foraging and the colony retreats into the nest.

The newly hatched bees look the same as those born in spring or summer, but their physiology tells a different story. Their abdomens resemble those of nurse bees, with ample fat reserves, while their thoraxes are equipped with the powerful flight

Ivy is one of the last flowers of the year, providing bees with crucial provisions for the winter.

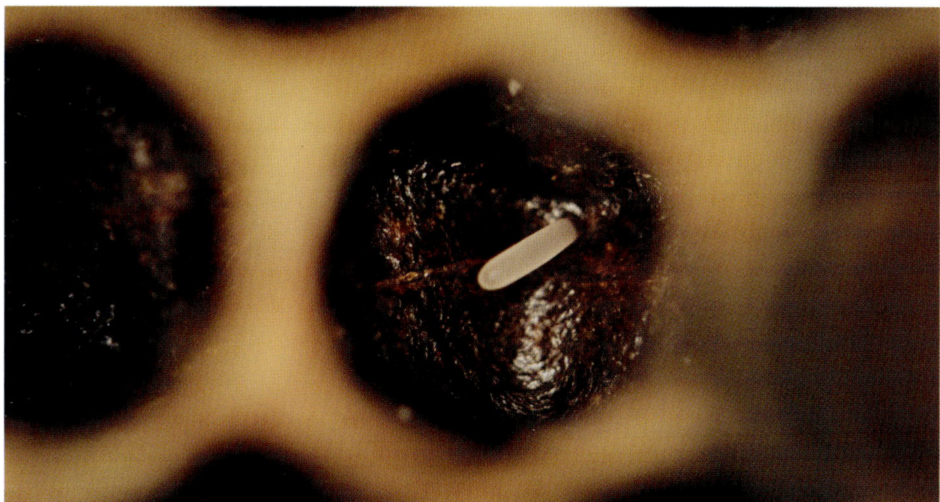

Eggs laid in the autumn develop into winter bees, which live until the following spring, spending nearly their entire lives inside the hive.

muscles of forager bees. This unique combination makes them ideal heater bees, capable of maintaining the temperature inside the nest. These winter bees possess another remarkable trait: They live up to four times longer than their summer-born sisters.

Despite sharing the same genetic code, winter bees exhibit vastly different characteristics. According to Prof. Jürgen Tautz, this can be explained by epigenetics. The slightly cooler brood temperature alters how the bees' genes are expressed, turning off some genes while turning on others. This phenomenon, though rare in nature, is also seen in crocodiles, where brood temperature determines whether males or females hatch. However, in bees, this mechanism is not randomly triggered but deliberate, allowing the colony to produce long-lived winter bees precisely when they are needed. These exceptional bees have one critical mission: to survive through winter and ensure the colony's renewal in spring. As soon as the temperature drops again, the winter bees cluster tightly together deep within the hive, conserving warmth and protecting the queen. Their unique physiology enables them to endure the harsh months, sustaining the colony until the warmth of spring returns. Soon, new eggs will be laid, and the next generation of bees will emerge. The colony perseveres, each bee playing its part in an endless cycle that balances the individual and the collective.

OVERLEAF: In late summer, bees often gather at the nest entrance to escape the heat inside.

About the authors

Dennis Wells holds a master's degree in sociology and has dedicated his career to exploring environmental issues. He has lived and studied in both Germany and Canada, completing parts of his education in Düsseldorf and Vancouver. After graduating, he taught at the University of Düsseldorf, worked in radio and eventually found his calling in documentary filmmaking. His films have won multiple international awards – including an Emmy for outstanding research, an AAAS Science Journalism Award and several honours at Germany's renowned nature film festival, Greenscreen. This non-fiction book is his first. He lives and works in Cologne and, together with his family, as close to the Rhine as possible.

Prof. Jürgen Tautz, a retired professor at the University of Würzburg, is a zoologist, sociobiologist and leading expert on honey bee behaviour. He is the author of numerous books, including the bestseller *The Buzz about Bees* (translated into 20 languages) and *Communication Between Honeybees*. His writing and popular lectures have twice been honoured by the European Molecular Biology Organization (EMBO), who included him among the best scientists in Europe in communicating science to the public. He has also received the prestigious Communicator Award of the German Research Foundation.

Acknowledgements

My heartfelt thanks go to everyone who worked on the film *Tagebuch einer Biene* (A Bee's Diary) because, without the film, this book would not exist. I am especially grateful to my fellow producers Bernd Wilting, Uli Veith and Niobe Thompson. The cinematography by Brian McClatchy, whose footage makes up most of the images in this book, played an outstanding role, as did the photographic work of Rudolf Diesel and Oliver Giel. The support of beekeepers Friedrich Benzenhöfer, Markus Graf and Roland Sachs was invaluable during filming, and I am particularly thankful to Prof. Tugrul Giray and Prof. Jürgen Tautz, who generously shared their expertise with me at every stage.

Picture credits

Further reading

Arndt, Ingo and Tautz, Jürgen. *Honey Bees.* The Natural History Museum, 2021.

Arndt, Ingo und Tautz, Jürgen. *Honigbienen: Geheimnisvolle Waldbewohner.* Knesebeck Verlag, 2020.

Chittka, Lars. *The Mind of a Bee.* Princeton University Press, 2022.

Menzel, Randolf und Eckoldt, Matthias. *Die Intelligenz der Bienen. Wie sie denken, planen, fühlen und was wir daraus lernen können.* Albrecht Knaus Verlag, 2019. (No English version available.)

Seeley, Thomas D. *Bienendemokratie.* S. Fischer Verlag, 2015.

Seeley, Thomas. *Honeybee Democracy.* Princeton University Press, 2010.

Tautz, Jürgen. *Communication Between Honeybees. More than Just a Dance in the Dark.* Springer International Publishing, 2022.

Tautz, Jürgen. *Phänomen Honigbiene.* Spektrum Akademischer Verlag, 2007.

Tautz, Jürgen. *The Buzz about Bees: Biology of a Superorganism.* Springer, 2008.

Tautz, Jürgen und Hülswitt, Tobias. *Das Einmaleins der Honigbiene: 66 x Wissen zum Mitreden und Weitererzählen.* Spektrum Akademischer Verlag, Heidelberg , 2019.

Tautz, Jürgen und Steen, Diedrich. *Die Honigfabrik. Die Wunderwelt der Bienen – eine Betriebsbesichtigung.* Gütersloher Verlagshaus, 2017.

Tautz, Jürgen and Steen, Diedrich. *The Honey Factory: Inside the Ingenious World of Bees.* Black Inc., 2018.

Tourneret, Eric, de Saint Pierre, Sylla and Tautz, Jurgen. *The Beeing: Life Inside a Honeybee Colony.* Deep Snow Press, 2021.

Index

Page numbers in *italic* refer to illustration captions and those in **bold** refer to feature boxes.

First published by the Natural History Museum,
Cromwell Road, London SW7 5BD. publishing@nhm.ac.uk.

Published in North America, South America, Central America and the
Caribbean by Smithsonian Books
PO Box 37012, MRC 513
Washington, DC 20013
smithsonianbooks.com

This book may be purchased for educational, business, or sales promotional use.
For information please write the Special Markets Department at the address or website above.

Hardcover ISBN: 978-1-58834-816-6
Library of Congress Cataloging-in-Publication Data
Names: Wells, Dennis (Film director) author | Tautz, Jürgen author
Title: Diary of a honey bee / Dennis Wells with Jürgen Tautz.
Description: London : Natural History Museum, [2025] | Includes bibliographical references and index.
Identifiers: LCCN 2025024839 | ISBN 9781588348166 hardcover
Subjects: LCSH: Honeybee--Life cycle | Honeybee--Behavior Honeybee--Anatomy
Classification: LCC QL568.A6 W45 2025 | DDC 595.79/9--dc23/eng/20250827
LC record available at https://lccn.loc.gov/2025024839

Printed in China, not at government expense
30 29 28 27 26 1 2 3 4 5

Cover: top (p.2), middle right (p.40), left of author (p.36), bottom right (p.70): Rudolf
Diesel/©taglichtmedia; top right (p.55), middle left (pp.118-119), bottom left (pp.146-147):
Brian McClatchy/©taglichtmedia.

Designed by Bobby Birchall, Bobby&Co, London
Reproduction by Saxon Digital Services
Printed by Toppan Leefung Printing Ltd., China
This book is printed using vegetable inks.
We hope you will continue to love this book, but when you have finished with it
please give to someone else or donate to charity.